BOYLE STUDIES

Boyle Studies

Aspects of the Life and Thought of Robert Boyle (1627–91)

MICHAEL HUNTER

Birkbeck, University of London, UK

Routledge
Taylor & Francis Group

LONDON AND NEW YORK

First published in paperback 2024

First published 2015 by Ashgate Publishing

Published 2012
by Routledge
4 Park Square, Milton Park, Abingdon, Oxon OX14 4RN

and by Routledge
605 Third Avenue, New York, NY 10158

Routledge is an imprint of the Taylor & Francis Group, an informa business

Copyright © Michael Hunter 2012, 2015, 2024

The right of Michael Hunter to be identified as author of this work has been asserted in accordance with sections 77 and 78 of the Copyright, Designs and Patents Act 1988.

Publisher's Note
The publisher has gone to great lengths to ensure the quality of this reprint but points out that some imperfections in the original copies may be apparent.

British Library Cataloguing in Publication Data
A catalogue record for this book is available from the British Library

Library of Congress Cataloging-in-Publication Data
Hunter, Michael Cyril William
    Boyle studies : aspects of the life and thought of Robert Boyle (1627–91) / by Michael Hunter.
        pages cm
    Includes bibliographical references and index.
    ISBN 978–1–4724–2810–3 (hardcover)

    1. Boyle, Robert, 1627–1691. 2. Scientists – Great Britain– Biography. 3. Nature – Early works to 1800. 4. Science – Early works to 1800.
    I. Title.
    Q143.B77H87 2015
    509.2–dc23
    [B]   2014026570

ISBN: 978-1-4724-2810-3 (hbk)
ISBN: 978-1-03-292442-7 (pbk)
ISBN: 978-1-315-56982-6 (ebk)

DOI: 10.4324/9781315569826

Contents

List of Plates

Acknowledgements

The following publishers of the books and journals in which studies included here first appeared have kindly given their permission for the material in question to be reprinted:

Chapter 1: Hitherto unpublished.

Chapter 2: Previously published as 'Robert Boyle's Early Intellectual Evolution: A Reappraisal', in special Boyle issue of *Intellectual History Review*, 25 (2015), 5–19. www.tandfonline.com.

Chapter 3: Previously published as 'Robert Boyle and the Early Royal Society: A Reciprocal Exchange in the Making of Baconian Science', *British Journal for the History of Science*, 40 (2007), 1–23 (reprinted by permission of Cambridge University Press).

Chapter 4: Previously published as 'Robert Boyle, Narcissus Marsh and the Anglo-Irish Intellectual Scene in the Late Seventeenth Century', in Muriel McCarthy and Ann Simmons (eds), *The Making of Marsh's Library: Learning, Politics and Religion in Ireland 1650–1750* (Dublin: Four Courts Press, 2004), pp. 51–75.

Chapter 5: Previously published in slightly abbreviated form as 'The Disquieted Mind in Casuistry and Natural Philosophy: Robert Boyle and Thomas Barlow', in Harald Braun and Edward Vallance (eds), *Contexts of Conscience in Early Modern Europe 1500–1700* (Basingstoke: Palgrave Macmillan, 2004), pp. 82–99 and 206–10.

Chapter 6: Previously published as 'Robert Boyle and Secrecy', in Elaine Leong and Alisha Rankin (eds), *Secrets and Knowledge in Medicine and Science 1500–1800* (Farnham: Ashgate, 2011), pp. 87–104.

Chapter 7: Previously published as 'Robert Boyle and the Uses of Print', in Danielle Westerhof (ed.), *The Alchemy of Medicine and Print: The Edward Worth Library, Dublin* (Dublin: Four Courts Press, 2010), pp. 110–24.

Chapter 8: Hitherto unpublished English version. French translation by Charles Ramond published as 'Boyle et le surnaturel', in Myriam Dennehy and Charles Ramond (eds), *La Philosophie Naturelle de Robert Boyle* (Paris: Librairie Philosophique J. Vrin, 2009), pp. 213–36. © Librairie Philosophique J. Vrin, Paris, 2009. http://www.vrin.fr.

Chapter 9: Hitherto unpublished.

In addition, the following acknowledgments should be recorded in relation to each individual chapter:

Chapter 1: The epilogue reuses material that appeared in an article entitled 'Genius Eclipsed: The Fate of Robert Boyle', *History Today*, November 2009, pp. 20–25; some of this is also to be found in my essay 'Newton's Style of Science', in Denis R. Alexander (ed.), *The Isaac Newton Guide Book* (Cambridge: Faraday Institute Publishing, 2012), pp. 27–33. For comments on this chapter, I am grateful to Peter Anstey, Michelle DiMeo and Lawrence M. Principe. For advice on the Appendix, I am grateful to Jim Bennett, Ted Davis, Vera Keller and Jack Macintosh.

Chapter 2: I am grateful to Peter Anstey, Elizabethanne Boran, David Cram and Michelle DiMeo for their careful reading of a draft of this chapter.

Chapter 3: This chapter is a revised version of a guest lecture given to the British Society for the History of Science at its extraordinary general meeting at the Royal Society on 8 June 2005. I am grateful to Janet Browne, Peter Bowler and Philip Crane for inviting me to give the lecture and for their assistance on that occasion. For comments on a draft, I am grateful to Peter Anstey, Mordechai Feingold, Guido Giglioni and Tina Malcolmson. Earlier versions of part of it were given (and have been cited) under different titles at the Second International Bacon Seminar at Queen Mary, University of London in September 2001, at Johns Hopkins University and Vanderbilt University in March 2002, and at a European Science Foundation workshop at the Herzog August Bibliothek, Wolfenbüttel in March 2004. I am grateful to Graham Rees, Lawrence M. Principe, Matthew Ramsay, Sachiko Kusukawa and Ian Maclean for hosting me on those occasions, and to the audiences for their comments. I have also benefited from comments by Harriet Knight, Daniel Carey and Ken Brown, and I am grateful to Joanna Corden and the archive staff at the Royal Society for their assistance.

Chapter 5: I am grateful to Martin Stone for discussions of this topic, and to Peter Anstey and various of the contributors to the volume in which it originally appeared for helpful comments on this chapter.

Chapter 6: I am grateful to Peter Anstey and Lawrence M. Principe for their comments on a draft of this chapter, to Victor Boantza for showing me the paper referred to in note 54 prior to publication, and to Will Poole for the further reference included in the same note.

Chapter 7: I am grateful to Bill McCormack and Elizabethanne Boran for inviting me give this paper at the Edward Worth Library in April 2008, and then to give it again at the conference to celebrate the 275th anniversary of the foundation of the library, held at the Royal Society in London in December that year. I would also like to record my profuse thanks to Dr Boran for introducing me to the Boyle items in the Worth Library and for providing digital images of them.

Chapter 8: This paper was delivered at the conference 'La philosophie naturelle de Robert Boyle' at the Université de Bordeaux III in March 2005, and I am grateful to those present for their comments.

Chapter 9: This paper was delivered at the conference 'New Worlds, New Philosophies' at Princeton University on 1 March 2012; it was subsequently given at Northumbria University on 2 October 2013. I am grateful to those present on both occasions for their comments. In addition, it has benefited from the advice of Peter Anstey, who also helped concerning the extract from the Locke MSS in Appendix 2, while Felicity Henderson, Joanna McManus and other members of the library and archive staff have kindly assisted my researches at the Royal Society.

Abbreviations

The following abbreviations are repeatedly used in this book:

Between God and Science	Michael Hunter, *Boyle: Between God and Science* (New Haven, CT and London, 2009)
Birch, *Royal Society*	Thomas Birch, *The History of the Royal Society of London* (4 vols, London, 1756–57)
BL	British Library
Boyle by Himself and His Friends	Michael Hunter (ed.), *Robert Boyle by Himself and His Friends* (London, 1994)
Boyle Papers	Michael Hunter, with contributions by Edward B. Davis, Harriet Knight, Charles Littleton and Lawrence M. Principe, *The Boyle Papers: Understanding the Manuscripts of Robert Boyle* (Aldershot, 2007)
BP	Royal Society Boyle Papers (volumes are denoted in the form 'BP 36')
Cl. P.	Royal Society Classified Papers
Correspondence	Michael Hunter, Antonio Clericuzio and Lawrence M. Principe (eds), *The Correspondence of Robert Boyle* (6 vols, London, 2001)
FRS	Fellow of the Royal Society
MS	Manuscript
NRRS	*Notes and Records of the Royal Society of London*
ODNB	*Oxford Dictionary of National Biography*
OED	*Oxford English Dictionary*
Phil. Trans.	*Philosophical Transactions*
QC	Queen's College, Oxford
RBC	Royal Society Copy Register Book
RS	Royal Society
Scrupulosity and Science	Michael Hunter, *Robert Boyle (1627–91): Scrupulosity and Science* (Woodbridge, 2000)
Workdiary	*The Workdiaries of Robert Boyle*, available at http://www.livesandletters.ac.uk/wd/index.html. References to this source throughout take

the form 'Workdiary 1–2' (that is, Workdiary number, followed by entry number)

Works　Michael Hunter and Edward B. Davis (eds), *The Works of Robert Boyle* (14 vols, London, 1999–2000)

Quotations from manuscript sources are presented according to the principles expounded in *Correspondence*, vol. 1, pp. xli–xlii, and *Works*, vol. 1, p. cii. Briefly, original spelling, capitalisation and punctuation are retained; standard contractions (for example, the thorn with superscript 'e' for 'the') have been silently expanded. Underlining in the original is shown by the use of italic. Original foliation or pagination has been indicated by the insertion in the text of 'fol. 2' or 'fol. 2v' within soliduses where each recto or verso of the manuscript text begins (or 'p. 2' where each page begins). Editorial insertions have been denoted by square brackets. Words or phrases inserted above the line in the original have been denoted <thus>. Deletions are recorded in footnotes or endnotes. However, in cases where a full transcription is already available in print, deletions and substitutions of words have generally been silently ignored.

Chapter 1

Introduction

Boyle in the Twenty-First Century

The status of Robert Boyle (1627–91) as the one of the greatest of English scientists has been a commonplace ever since his own time. But our appreciation of him has undoubtedly been significantly enhanced over the past thirty years. Prior to that, Boyle was lionised but rather taken for granted, and it is striking that, in surveys of seventeenth-century science written in the 1960s and 1970s, he does not always occupy a very prominent role.[1] Yet the state of affairs has now changed completely, and Boyle's position as the most influential English natural philosopher in the generation before Isaac Newton is today generally acknowledged. Indeed, in a recent survey of the philosophy of the period, a whole chapter is devoted to him.[2] For this, various reasons may be given.

One has undoubtedly been the increasing availability of source material relating to Boyle. Scholars had long been aware that a vast collection of his papers and letters survived at the Royal Society, having been bequeathed to that institution in 1769 and roughly sorted and bound in the 1850s: in total, the archive comprises over 20,000 leaves bound in some eighty volumes. Tentative investigations of this resource were made by a few scholars in the 1950s and 1960s, but at that point the collection remained (in the words of one commentator) 'uncatalogued and uncataloguable'.[3] In 1986, however, I undertook the task of cataloguing of the archive, and in 1992 a printed catalogue was issued in conjunction with a complete microfilm of the collection which made it widely accessible for the first time.[4] A revised version of the catalogue was issued in 2007, by which time the core volumes of the archive had been made available

---

[1] See, for example, A.G.R. Smith, *Science and Society in the Sixteenth and Seventeenth Centuries* (London, 1972), or Robert Mandrou, *From Humanism to Science 1480–1700* (1973; English edn, Harmondsworth, 1978).

[2] See Peter Anstey (ed.), *The Oxford Handbook of British Philosophy in the Seventeenth Century* (Oxford, 2013), ch. 3.

[3] M.A. Stewart, 'The Authenticity of Robert Boyle's Anonymous Writings on Reason', *Bodleian Library Quarterly*, 10 (1978–82), 280–89, on p. 283. For a history of the archive, see *Boyle Papers*, pp. 22ff.; on its early exploitation, especially by R.S. Westfall, M.B. Hall and R.E.W. Maddison, see ibid., p. 29.

[4] See Michael Hunter, *Letters and Papers of Robert Boyle: A Guide to the Manuscripts and Microfilm* (Bethesda, MD, 1992).

online: the updated catalogue formed the central component of a volume entitled *The Boyle Papers: Understanding the Manuscripts of Robert Boyle*, in which it was accompanied by various interpretative studies of the archive which further contributed to its elucidation.[5]

A subsequent, key development was the preparation and publication of definitive editions of Boyle's *Works* in 14 volumes in 1999–2000 and his *Correspondence* in six volumes in 2001. The former was edited by me in conjunction with Edward B. Davis, and the latter by a team comprising myself, Antonio Clericuzio and Lawrence M. Principe, and it could be argued that the availability of these resources had a transformative effect on studies of Boyle. As Peter Dear strikingly put it at the start of his review of the former: 'This monumental piece of scholarship takes its place on the shelf alongside other classic "complete works" editions of great men, such as the Adam and Tannery Descartes or Favaro's Galileo', and the *Correspondence* has been the subject of comparable evaluations.[6]

Meanwhile, attention to Boyle had received a fillip from a rather different source in the form of two publications: Steven Shapin and Simon Schaffer's *Leviathan and the Air-pump: Hobbes, Boyle and the Experimental Life* (1985) and Steven Shapin's *A Social History of Truth: Civility and Science in Seventeenth-century England* (1994). It would not be an exaggeration to say that these two books have reshaped the study of early modern science (and of the history of science as a whole), and Boyle plays a central role in both, thus giving him a greater prominence in the historiography than ever before. *Leviathan and the Air-pump* focuses on the controversy between Boyle and Thomas Hobbes over the interpretation of Boyle's findings with his air-pump, and the theme that proved most influential was the book's insistence on Boyle's stress on the establishment of reliable 'matters of fact' as the key to intellectual endeavour, the purpose being (it was claimed) to undercut more hypotheticalist and *a priori* approaches that were seen as intellectually and socially disruptive. With *A Social History of Truth*, the emphasis was on social status as the chief criterion by which credibility was assessed, and Boyle was presented as the prime model for the conduct of intellectual life, his credentials being effortlessly established through a combination of gentility and probity which is explored at length in that book.

---

[5]   *The Boyle Papers* was a collaborative work in which my co-authors were Edward B. Davis, Harriet Knight, Charles Littleton and Lawrence M. Principe. For the online material, see http://www.bbk.ac.uk/boyle/boyle_papers/boylepapers_index.htm (but see below, n. 18).

[6]   Peter Dear, 'Essay Review: Boyle in the Bag!', *British Journal for the History of Science*, 35 (2002), 335–40, on p. 335; for the *Correspondence*, see especially Noel Malcolm, 'Of Air and Alchemy', *Times Literary Supplement*, 23 August 2002, pp. 10–11.

These tropes have now become predominant in interpretations of the science of Boyle's period, and interest in Boyle has been enhanced accordingly.[7]

Yet it is ironic that these 'constructivist' readings of Boyle arguably represented an ingenious gloss on the image of him as an exemplary scientist that had been prevalent since his own time, enshrined particularly in the earliest and most influential biography of him by the cleric and antiquary Thomas Birch, which was prefixed to his 1744 edition of Boyle's works. By contrast, the 1980s and 1990s saw a complete revaluation of Boyle, based largely on the exploitation of material in his archive. One important facet of this was the discovery that he was much more interested in alchemy than had previously been realised, as has been illustrated particularly by Lawrence M. Principe in his magisterial study *The Aspiring Adept: Robert Boyle and His Alchemical Quest* (1998).[8] Alchemy had been particularly problematic for Birch, who wanted to make Boyle seem as rationalist and modern as possible and who went out of his way to suppress evidence of such interests on his hero's part; but it is striking how echoes of a similar attitude are to be found in *Leviathan and the Air-pump*.[9] In parallel with this, I had explored Boyle's spiritual and mental life, illustrating that, in contrast to the benign and purposeful image painted by Birch, his religious life was in fact stressful and complicated.

Even in the case of alchemy, Boyle had reservations about its study, not because it might be false, but because it might be illicit – that it might tempt the unwary to try to control nature through a league, not with God, but with his great adversary, the Devil, thus endangering their immortal soul.[10] This was just one of many dilemmas that Boyle encountered during his assiduous examination of his conscience, the records of which have proved crucial to a fuller understanding of his inner mental life. Perhaps the most striking example of these comprises a document surviving among the Boyle Papers which vividly records how Boyle

---

[7]   For an appreciative appraisal, see Jan Golinski, *Making Natural Knowledge: Constructivism and the History of Science* (Cambridge, 1998; new edn, 2005); for a more critical view, see Michael Hunter, 'Scientific Change: Its Setting and Stimuli', in Barry Coward (ed.), *A Companion to Stuart Britain* (Oxford, 2003), pp. 214–29.

[8]   Lawrence M. Principe, *The Aspiring Adept* (Princeton, NJ, 1998). See also his important study with William R. Newman, *Alchemy Tried in the Fire: Starkey, Boyle and the Fate of Helmontian Chymistry* (Chicago, IL, 2002).

[9]   For Birch, see Michael Hunter, 'Robert Boyle and the Dilemma of Biography in the Age of the Scientific Revolution', in Michael Shortland and Richard Yeo (eds), *Telling Lives in Science: Essays on Scientific Biography* (Cambridge, 1996), pp. 115–37, esp. pp. 129–33, reprinted in *Scrupulosity and Science*, ch. 11, esp. pp. 264–6; for *Leviathan and the Air-pump*, see Principe, *Aspiring Adept*, esp. pp. 107–9, 111.

[10]   See Michael Hunter 'Alchemy, Magic and Moralism in the Thought of Robert Boyle', *British Journal for the History of Science*, 23 (1990), 387–410, reprinted in *Scrupulosity and Science*, ch. 5.

suffered from religious doubts and how he agonised over moral decisions that he made, for instance concerning the amount of his massive inherited income that he should devote to charity.[11] From such texts, Boyle comes across as an anxious and mixed-up figure – in complete contrast to the suave and controlled image conveyed by Birch's *Life*, even if perhaps more to the taste of the twenty-first century. Moreover, it can be argued that this was integrally connected to his career as a scientist, since (as he told Bishop Burnet, one of those who advised him on the scruples that affected him) 'he made Conscience of great exactnes in Experiments'.[12]

If Boyle's tortured spiritual life is one aspect of the 'new' Boyle brought to light by archival studies, the other is a surprising degree of convolution in his dealings with the wider world – again, in contrast to the traditional image of him as assured and decisive. A key example of this is provided by Boyle's views on medicine and his relations with the medical profession of his day. He seems to have come to the conclusion that the therapy that was commonly purveyed was almost worthless and needed to be replaced by a new, scientific medicine based on empirical principles. Yet, although he wrote a book expounding such views, in the end he suppressed it, not least because of his fear of offending doctors.[13] Also revealing is his indecision about publishing a collection of medical recipes, despite the obvious benefit which their dissemination would have had for those in need of health care. Again, he was worried about treading on the toes of medical men, but he was also concerned about the reliability of the recipes, since he certainly did not want to purvey nostrums that were dangerous; in addition, he wondered whether he should divulge the names of those from whom he had obtained the recipes, in both cases devising convoluted strategies for resolving such dilemmas.[14] The perhaps surprising pusillanimity which he showed in this connection can be matched elsewhere, for instance in his apologies for the

---

[11] See 'Casuistry in Action: Robert Boyle's Confessional Interviews with Gilbert Burnet and Edward Stillingfleet, 1691', *Journal of Ecclesiastical History*, 44 (1993), 80–98, reprinted in slightly extended form in *Scrupulosity and Science*, ch. 4.

[12] See 'The Conscience of Robert Boyle: Functionalism, "Dysfunctionalism" and the Task of Historical Understanding', in J.V. Field and F.A.J.L. James (eds), *Renaissance and Revolution: Humanists, Scholars, Craftsmen and Natural Philosophers in Early Modern Europe* (Cambridge, 1993), pp. 147–59, on p. 158, reprinted in *Scrupulosity and Science*, ch. 3, on p. 69.

[13] See 'Boyle versus the Galenists: A Suppressed Critique of Seventeenth-century Medical Practice and its Significance', *Medical History*, 47 (1997), 322–61, reprinted in *Scrupulosity and Science*, ch. 8.

[14] See 'The Reluctant Philanthropist: Robert Boyle and the "Communication of Secrets and Receits in Physick"', in O.P. Grell and Andrew Cunningham (eds), *Religio Medici: Medicine and Religion in Seventeenth-century England* (Aldershot, 1996), pp. 247–72, reprinted in *Scrupulosity and Science*, ch. 9.

defects of his published books and his slightly paranoid anxiety that his ideas might be stolen by others.[15] Here too, Boyle's relations with his milieu were less serene and effective than has sometimes appeared.

I presented this new view of Boyle as a rather convoluted figure in various conference papers and journal articles that were then collected together as a book, *Robert Boyle (1627–91): Scrupulosity and Science* (2000). Since then, I have produced a more synthetic biography, *Boyle: Between God and Science* (2009), which seeks to do justice to Boyle's overall role as a founding father of modern science, combining these revisionist findings with an emphasis on the more positive facets of his achievement, particularly his championship of controlled experiment and of the mechanical philosophy, and his sophisticated reflections on the epistemology of science and the relations between God and nature. But it could be argued that that synthesis was only possible because I had already executed the more detailed research concerning hitherto unknown aspects of Boyle's life and thought that was presented in my 2000 volume. Moreover, since publishing that, I have continued to make similar studies, and it is these that the current work comprises – many of them already published, but one appearing here for the first time. As with the book published in 2000, readers will here find detailed expositions of important themes that were summarised only briefly, or in some cases not dealt with at all, in *Boyle: Between God and Science*. This book thus constitutes an important supplement to that one, which should also be of interest in its own right.

The Current Volume

We start with a chapter which vindicates the claim made in my 1995 article 'How Boyle Became a Scientist' (reprinted in *Robert Boyle: Scrupulosity and Science*) that Boyle only discovered science in 1649 after previously being preoccupied by ethical and literary concerns. By way of background, it should be explained that this claim derived directly from the intensive study of the Boyle Papers that has already been referred to: whereas previously that archive had tended to be regarded as an unvariegated mass, the use of handwriting and other evidence made it possible to discern distinct strata within it, and one such stratum turned out to comprise the writings of Boyle's early scientific phase, which could be distinguished from the very different, moralistic one which preceded it. Also relevant was a further outcome of the full exploitation of the archive, namely the recognition that a crucial set of documents existed within it which have been christened Boyle's 'workdiaries' and have now been organised chronologically

---

[15] See ibid., ch. 7.

and published in online form.[16] These comprise a series of paperbooks in which he recorded data of various kinds, and their significance here is that, at just this point, their content changed dramatically from moral apothegms to the recording of recipes and comparable material. It was on the basis of this and other archival evidence that I claimed that Boyle's preoccupations changed dramatically in his early twenties. Although some have preferred to retain a view of Boyle as being a scientist almost from the outset, the restatement of my case presented here makes this increasingly problematic. Chapter 2 also reflects more broadly on the historiography of Boyle's early years, considering the 'origins stories' about him that have circulated both in his own time and since, evaluating his relations with the groups with which he came into contact and stressing the overriding intellectual eclecticism on his part that helps to explain his significance.

Chapter 3 offers a further reconsideration of Boyle's relations with his context, in this case in connection with the early Royal Society. Boyle's close links with that institution in its formative years are legendary, and the society is often seen as putting into practice a programme pioneered by Boyle. Here, on the other hand, it is suggested that their relationship may have been a more reciprocal one than has often been thought – a claim again derived from an intensive study of Boyle's corpus, in this case not only his manuscripts, but also his printed works. This revealed an important development in Boyle's natural philosophical method from the 1660s onwards: his use of 'heads' and 'inquiries' as a means of organising his data, setting himself an agenda when studying a subject, and soliciting information from others. Boyle acknowledged that he derived this approach from his mentor, Francis Bacon (1561–1626), but he had not previously used it in his work, and the reason why it came to the fore when it did is not apparent from his own corpus.

Here, it is argued that the cause of this crucial methodological change on Boyle's part was the influence on him of the Royal Society, which was characterised in its early years by a concern for systematic data-collecting from which, in this case, Boyle seems to have learned, rather than vice versa. In this connection, it is shown that a key text, Boyle's influential 'General Heads for a *Natural History of a Country*, Great or small', published in *Philosophical Transactions* in 1666, represents more of a shared initiative between him and the society than has hitherto been appreciated (and, revealingly, the society's Secretary, Henry Oldenburg, seems to have felt it quite appropriate to produce an adapted version of it, which is printed as an appendix to the chapter). Hence, this chapter throws significant light on this aspect of the society's work in its own

---

16   See http://www.livesandletters.ac.uk/wd/index.html. For a full account, see Michael Hunter and Charles Littleton, 'The Workdiaries of Robert Boyle: A Newly Discovered Source and its Internet Publication', *NRRS*, 55 (2001), 373–90, reprinted in extended form in *Boyle Papers*, ch. 3.

right, while also exemplifying how external factors often need to be taken into account in order fully to understand changes evidenced in Boyle's own writings.

Chapter 4 offers a rather different view of Boyle's relations with his milieu, in this case taking the form of a survey of his Irish links; it was originally presented at a conference marking the tercentenary of the foundation of Archbishop Marsh's library in Dublin in 2001. Though it ranges more widely, the chapter focuses particularly on initiatives with which Boyle was associated in the 1680s, notably the project for publishing the Bible in Irish and the scientific developments that form the background to the foundation of the Dublin Philosophical Society in 1683. In both cases, Boyle was closely associated with Narcissus Marsh, Provost of Trinity College Dublin from 1679 onwards, and attention is paid to the shared concerns of the two men, and also to the Oxford background from which Marsh emanated and with which both he and Boyle retained close contacts. In the course of the chapter, further light is thrown on Boyle's dilemmas of conscience, in this case concerning the impropriations on church lands in Ireland that he was granted as part of the Restoration settlement, which can in part be seen as providing the background to his involvement in the Bible project, though that also reflected his enthusiasm for such evangelistic initiatives more generally. On the other hand, the Bible project proved not to be without problems in its own right. Equally interesting is the evidence that is provided of Boyle's interest in the comets of 1680–82 and the observations of them made by Marsh and others, and of his association with Dublin virtuosi who visited London, such as Allen Mullen, both of which illustrate the breadth of his natural philosophical concerns.

The next three chapters provide further evidence of the 'convoluted' Boyle divulged in *Robert Boyle: Scrupulosity and Science*. Chapter 5 goes to the heart of Boyle's tortured spiritual life, providing an account of a group of manuscripts dealing with cases of conscience that were explicitly written at his behest but which survive, not in Boyle's own archive, but among the papers of his spiritual advisor, Thomas Barlow, now at Queen's College, Oxford. These fascinating, if complex, documents were written by Barlow in the 1680s, by which time he had become Bishop of Lincoln. In them, he examined some of the more difficult issues of casuistry by which Boyle seems to have been preoccupied, especially concerning the role of 'scruples' and the criteria according to which decisions on matters of conscience should properly be made. By Barlow's own admission, such topics rarely formed the subject of self-contained disquisitions of the kind that he wrote for Boyle, but they nevertheless resonated with his pastoral experience, from which various fascinating examples were included by way of illustration; the chapter therefore sheds interesting broader light on the concern with practical divinity of churchmen like Barlow as well as providing new evidence concerning the preoccupations of Boyle himself.

In Boyle's case, it is clear that the main problem that faced him was the difficulty of reaching clear decisions on moral issues of the kind with which casuistry was concerned, and this interestingly echoes the equivocation in his philosophical writings for which he was notorious. This thus raises the issue of the link between Boyle's casuistry and his natural philosophy, which was one of the topics considered in *Robert Boyle: Scrupulosity and Science* and which is further appraised in this chapter. As to the extent to which Barlow was able to assist Boyle, it seems likely that, though Barlow's intellectual outlook was rather different from Boyle's, the bishop was able to offer a clarity and rigour in his analysis of the issues with which he dealt that was almost certainly comforting to Boyle. On the other hand, it seems likely that, ultimately, neither his nor anyone else's advice could do much to quell the persistent prevarication which seems to have characterised Boyle's emotional and intellectual life.

Chapter 6 examines Boyle's attitude to secrecy, which turns out to be typically complex. His statements on the subject, which are analysed here, provided potential justification for a variety of practices: these ranged from the complete withholding of information of the kind that was typical of many craft traditions of the day and which was also often echoed in virtuoso values, through a kind of strategic use of secrecy as a bargaining counter, to a total openness that Boyle often seems to have considered proper to science. In practice, Boyle *did* often seek to demystify or disseminate information that had previously been treated as secret by those who possessed it. But on other occasions he prevaricated, using a variety of means to withhold all or part of information that he might usefully have divulged, and this occurred not only in predictable contexts like alchemy, but even in relation to the Royal Society: Boyle was behind various initiatives aimed at taking advantage of the institutional authority of that body to seal up and preserve information for the benefit of its originator. Indeed, there is some evidence that contemporaries found Boyle's attitudes frustratingly inconsistent, and this raises issues about his image to which we will return later in this introduction.

Boyle's attitude towards the medium of print also proves to be revealingly ambivalent, as Chapter 7 illustrates. Boyle published prolifically in the course of his career, and he had a high view of the role of the printing press in giving wide circulation to knowledge. On the other hand, he was capable at times of using print in more strategic ways, for instance by having material privately printed for limited circulation. Also notable is the extent to which he used the front and back matter of his books – and sometimes also special publications – to apologise for the defects of his works, to assert his rights as an author and to trumpet his concern that he was the subject of intellectual theft or plagiarism. The result was to illustrate a revealing insecurity which echoes the querulousness on Boyle's part documented in various chapters of *Robert Boyle: Scrupulosity and Science*, and which needs to be taken into account if we are properly to understand him.

Finally, we have two chapters, one hitherto published only in French and the other not at all, which throw important light on topics that preoccupied Boyle in the last few years of his life. At that point, he took great interest in collecting authenticated examples of what he described as 'supernatural' phenomena: he intended to publish these in a work entitled 'Strange Reports', which would have had the apologetic purpose of providing empirical proof of a realm above the purely natural, thus disarming potential religious sceptics. Chapter 8 places this in context by considering Boyle's attitude to natural and supernatural causation more generally. He did indeed make a clear distinction between the two, and was critical of those who invoked supernatural causes when purely natural ones were adequate. But he was equally insistent that supernatural phenomena really existed and were potentially verifiable by empirical means, and it was to this end that he recorded a number of bizarre stories that would have appeared in his planned collection. On the other hand, Boyle withheld the 'supernatural' part of 'Strange Reports' from the only volume that appeared before his death of the work of which it would have formed part, *Experimenta & Observationes Physicæ*, invoking reasons of 'discretion'. Hence, once again, it is only through archival research that his intentions have come to light.

Lastly, Chapter 9, the most substantial and perhaps the most significant single component of this volume, throws new light on Boyle's interest in far-flung parts of the world. It does so by providing an account of another compilation on which Boyle worked in his later years, again an intended component of *Experimenta & Observationes Physicæ* that failed to come to fruition due to his death. In this case, the work was to have been entitled 'Physica Peregrinans, or the Travelling Naturalist', and it would have comprised a collection of accounts of exotic phenomena with which he was provided by travellers and others, many of them recorded in his workdiaries. Its objectives were twofold. In part, it had the polemical aim of illustrating the bankruptcy of traditional approaches to nature by showing how, in contrast to the rationalisation of common experience central to the scholastic natural philosophy of his day, nature was often surprising and exciting in its potency and variety. People's conception of what was possible therefore needed to be expanded accordingly. But, in addition, Boyle seems to have seen the material that he collected as a textual resource which was entertaining and educative in its own right, perhaps comparable to the romances that he had enjoyed as an adolescent. This point is illustrated by reference to Boyle's 'The Aspireing Naturalist', a romance set in a distant island, the hitherto unpublished surviving fragment of which is included as an appendix to Chapter 9. It is further borne out by a second appendix comprising a selection of hitherto unpublished narratives of the kind that 'Physica Peregrinans' was to have comprised, which shows just how vivid and absorbing these are. This concluding chapter thus throws light on a hitherto unknown facet of Boyle's intellectual agenda, acting as a suitable climax to the book as a whole.

The Way Forward: Boyle in Context

The material presented here offers detailed elucidation of various of Boyle's preoccupations, contributing to the reassessment of him that has been in progress for the last thirty years and to which many scholars have contributed.[17] This revaluation continues, as is exemplified by studies of different facets of Boyle that have recently appeared, ranging from Dmitri Levitin's re-examination of Boyle's views on the history of ancient philosophy to Richard Yeo's investigation of Boyle's intellectual method, demonstrating the extent to which he relied on his memory to supplement his voluminous notes.[18] There can be no doubt that there is more to be said about almost every aspect of Boyle's thought, and it seems likely that the revaluation of him that has occurred in recent years will in retrospect be seen as merely the beginning of an ongoing tradition.

Here, it is worth pointing out the extent to which the Boyle archive itself, already so much in evidence in the studies that follow, has potential to provide further insights into Boyle. Indeed, for all the exploitation of this resource that has already occurred – documented in the revised version of the catalogue of it published in *The Boyle Papers* (2007) – there are undoubtedly further riches to be discovered in this vast and multi-faceted deposit. In particular, it seems likely that much miscellaneous material is present because it was collected by Boyle himself, and that it may therefore have a significance that has hitherto been overlooked. Obviously caution is needed here, since it is clear that extraneous material was added to the Boyle Papers after Boyle's death due to the conflation of papers belonging to its early custodians, notably John Warr and Henry Miles.[19] On the other hand, much of the miscellaneous material that dates from Boyle's own lifetime may well have been there from the outset. Hence, through

---

[17] For a volume which forms a kind of anthology of revisionist approaches to Boyle, based on a conference held at Stalbridge in Dorset in December 1991 to mark the tercentenary of his death, see Michael Hunter (ed.), *Robert Boyle Reconsidered* (Cambridge, 1994). Many of the contributors to this subsequently went on to write book-length studies of different aspects of Boyle's thought.

[18] Dmitri Levitin, 'The Experimentalist as Humanist: Robert Boyle on the History of Philosophy', *Annals of Science*, 71 (2014), 149–82; Richard Yeo, 'Loose Notes and Capacious Memory: Robert Boyle's Note-taking and its Rationale', *Intellectual History Review*, 20 (2010), 335–54, and Richard Yeo, *Notebooks, English Virtuosi and Early Modern Science* (Chicago, IL, 2014), chs 5–6. For a bibliographical essay giving a comprehensive account of the literature on Boyle up to 2009, see *Between God and Science*, pp. 257–90. A more complete listing of writings on Boyle in alphabetical order is to be found on the Boyle website, http://www.bbk.ac.uk/boyle/researchers/boyle-bibliography.htm. It is planned that the Boyle website will move to the Royal Society in the near future, and that it will continue to be kept up to date at its new home.

[19] For material that may be intrusive, see *Boyle Papers*, esp. pp. 50, 60, 64.

such items, we may gain insight into matters in which he was interested but on which he rarely expressed an opinion – including the broad political and religious issues of his day on which he was notoriously evasive. An example of this is provided by the otherwise unknown draft legislative documents relating to slavery that have come to light in the archive and were published by Ruth Paley, Cristina Malcolmson and myself in an article in *Slavery and Abolition* in 2010.[20] Comparable items include religious and political documents in Volumes 3, 4 and 40–42 of the Boyle Papers, while it is also worth drawing attention to the copies in the archive of millenarian calculations and verse prophecies concerning political events, often in the hands of Boyle's favoured amanuenses, thus illustrating an interest on Boyle's part in material of a kind that some might find surprising.[21]

Of the political items, among the most striking are copies of the speeches before execution of Elizabeth Gaunt and John Ayloffe, both implicated in the Rye House Plot in 1683, and again both in the hands of trusted amanuenses.[22] There is also evidence that documents relating to the Popish Plot once formed part of the archive although they are now lost, having evidently been disposed of by Henry Miles.[23] These may evidence a greater sympathy by Boyle with political dissidents than has hitherto been recognised, thus mirroring the empathy with religious radicals that has already been shown to be part of his slightly idiosyncratic religious outlook.[24] On the other hand, we may here see evidence concerning another crucial matter that requires attention, namely a fuller understanding of Boyle's immediate milieu and the likely implications of this for his intellectual outlook. For we need to remember that these items date from the period when Boyle's shared a house in Pall Mall with his sister, Katherine, Lady Ranelagh, and it may well be that we here have clues concerning the political agenda of the joint household. Indeed, this also applies to the millenarian documents already referred to, since in his diary entry for 18 June 1690, John Evelyn specifically notes how he went with William Lloyd, Bishop of St Asaph, to the Pall Mall house, and that it was to 'Mr Boyle & Lady Ranelagh

---

[20]   See Ruth Paley, Cristina Malcolmson and Michael Hunter, 'Parliament and Slavery, 1660–c.1710', *Slavery and Abolition*, 31 (2010), 257–81.

[21]   For the latter, see esp. BP 3, fols 98–9; BP 41, fols 87–8, 97–9, 118; BP 42, fols 226, 254–5. See also Royal Society MS 189, fol. 2.

[22]   BP 38, fol. 75; BP 41, fols 119–22. The former comprises the latter part only of Gaunt's speech: see *Mrs Elizabeth Gaunt's Last Speech* [London, 1685], from which the BP text has minor verbal variants. The latter is apparently unpublished, as are the letters that accompany it: see *ODNB* and G. de F. Lord, 'Satire and Sedition: The Life and Work of John Ayloffe', *Huntington Library Quarterly*, 29 (1966), 255–73.

[23]   See *Boyle Papers*, pp. 83, 109.

[24]   See *Scrupulosity and Science*, pp. 56–7.

his sister' that Lloyd 'explained the necessity of its so falling out by the Scriptures in a very wonderfull manner, which he most skillfully & learnedly made out'.[25]

In the past few years, Lady Ranelagh has begun to emerge from the shadows as one of the most remarkable women in seventeenth-century England, far more significant in her own right than simply as the house-mate of her eminent brother. Recent studies have demonstrated her bold stance as a political commentator, never shy of blunt appraisals of the current state of affairs, usually from a strongly religious and moralistic viewpoint, and well enough connected to have significant influence on those in positions of authority.[26] As the intelligencer Samuel Hartlib nicely put it in 1659, she was a woman of 'piety and parts, with power and interest, which she hath and will always have with all the great ones, let them change never so often', and it is clear that she continued to play a role similar to that which she had had in the Interregnum during the period when she and Boyle lived together.[27] In many ways, it could be argued that her concern was always to assert the primacy of morality over politics, which made her comments slightly disturbing to seasoned politicians like the Earl of Clarendon. On the other hand, it is easy to understand how, for precisely this reason, her political stance was one that would have appealed to Boyle, since it would have resonated with the moralism that has come to the fore in our understanding of him. In spite of his own aversion to direct involvement in politics or government, it seems highly likely that he would have admired and supported his sister's stance.

Moreover, there was an important area where the siblings' concerns overlapped, namely medicine, and here it may well be possible to discern a joint stance that has hitherto been overlooked. Lady Ranelagh had taken a strong interest in medical and chemical matters in the 1640s and 1650s, as is documented by references to her both in the Hartlib Papers and in Boyle's own writings, where she is credited particularly with playing an active role in the cure of rickets.[28] She clearly continued to take an interest in such matters thereafter

---

[25]   Evelyn, *Diary*, ed. E.S. de Beer (6 vols, Oxford, 1955), vol. 5, pp. 25–6. What seems to be Boyle's record of the same meeting is in RS MS 187, fols 141–2, though with the date of 22 June. Cf. also Evelyn, *Diary*, vol. 4, pp. 636–7. For Boyle's and Lady Ranelagh's earlier millenarian interests, see *Between God and Science*, p. 86.

[26]   See Ruth Connolly, '"A Wise and Godly Sybilla": Viscountess Ranelagh and the Politics of International Protestantism', in Sylvia Brown (ed.), *Women, Gender and Radical Religion in Early Modern Europe* (Leiden, 2007), pp. 285–306; Ruth Connolly, 'A Proselytising Protestant Commonwealth: The Religious and Political Ideals of Katherine Jones, Viscountess Ranelagh' (1614–91)', *The Seventeenth Century*, 23 (2008), 244–64; Carol Pal, *Republic of Women: Rethinking the Republic of Letters in the Seventeenth Century* (Cambridge, 2012), ch. 5.

[27]   Quoted in ibid., p. 170.

[28]   See Michelle DiMeo, '"Such a Sister became such a Brother": Lady Ranelagh's Influence on Robert Boyle', *Intellectual History Review*, 25 (2015), 21–36, and Michelle

and was supportive of Boyle's experimental activities after he moved into her house at Pall Mall, even arranging for his laboratory facilities to be extended when alterations were made to the house in the 1670s.[29] It also seems likely that she was at least partly responsible for what would otherwise be an uncharacteristic initiative on Boyle's part, about which he suffered typical convolution that has already been noted, namely his composition of 'Considerations & Doubts Touching the Vulgar Method of Physick'.[30] Arguably, what needs to be explained is not so much why Boyle had second thoughts about this work, but why he wrote it in the first place, and here it seems plausible that we see the influence of the anti-establishmentarian and moralistic attitudes of his sister. Indeed, though this possibility has not hitherto been canvassed, current work by Michelle DiMeo has revealed verbal echoes between the extant texts of Boyle's anti-Galenist manifesto and the medical writings of Lady Ranelagh which imply that she may have played a significant role in prompting him to this uncharacteristically confrontational work.[31]

If Lady Ranelagh was the most significant figure living under the same roof as Boyle, there were also various others in the form of the assistants and amanuenses who helped him in writing his books and carrying out his experiments. These, too, clearly formed a significant part of Boyle's immediate intellectual milieu about which we ought to know more. Boyle rarely says much about these figures in his published writings, and on this basis they have been characterised as his 'invisible technicians'.[32] In fact, however, it is possible to identify a number of them, and in some cases to speculate about their mutual relations with Boyle. A few went on to become significant natural philosophers in their own right, most famously Robert Hooke, but also Denis Papin and Frederic Slare, all of whom have now been the subject of substantial scholarly study.[33] The case of Ambrose

---

DiMeo, 'Katherine Jones, Lady Ranelagh (1615–91): Science and Medicine in a Seventeenth-century Englishwoman's Writing' (PhD thesis, University of Warwick, 2009).

[29] See *Between God and Science*, pp. 166–7 and 332 n. 15.

[30] See above, p. 4.

[31] I am grateful to Michelle DiMeo for discussions of her findings, which she will divulge in a future publication.

[32] Steven Shapin, 'The Invisible Technician', *American Scientist*, 77 (1989), 554–63; see also Shapin, *Social History of Truth*, ch. 8.

[33] On Hooke, the literature is now voluminous: for a complete bibliography up to 2006, see Michael Cooper and Michael Hunter (eds), *Robert Hooke: Tercentennial Studies* (Aldershot, 2006), pp. 259–98. On Slare, see M.B. Hall, 'Frederic Slare, F.R.S. (1648–1727)', *NRRS*, 46 (1992), 23–41. On Papin, see esp. David Wootton, *The Scientific Revolution: The Invention of Science 1500–1750* (London, forthcoming), ch. 15, and the references there cited.

Godfrey Hanckwitz, who assisted Boyle in the making of phosphorus and went on to a career in related activities, has also attracted attention.[34]

But only recently has it come to light that one of Boyle's assistants was Thomas Emes, who was to achieve notoriety in early eighteenth-century London in the episode concerning the French Prophets, when it was predicted that he would rise from the dead but he failed to do so.[35] Previously, Emes had published tracts in a dispute about the fashionable acid-alkali theory, one of which has been described as 'a quite excellent summary of the intellectual issues in the medical debate'; in the course of these, he divulged that during the time that he had worked for Boyle, he had assisted him in his experiments on human blood, and this must have acted as an apprenticeship for him that he was later able to put to good use.[36] Another former assistant who became involved in controversy following Boyle's death was Robert St Clair, who published a critique of Thomas Burnet's famous *Sacred Theory of the Earth* in the course of which he, too, revealed his links with Boyle, advocating in a Boylean manner that nature should be studied by 'Observation and Experience', attacking Cartesian and other system-building of the kind that he saw Burnet as exemplifying, and arguing that Burnet's theory was incompatible with scripture.[37]

St Clair also referred to chemical reactions, including ones involving phosphorus that he had observed when he lived with Boyle, using such data as the basis for an alternative view of subterranean activity to that offered by Burnet, and a similarly idiosyncratic use of Boyle's and related ideas is exemplified by another of his employees, Hugh Greg. Greg was one of Boyle's most trusted amanuenses in his later years, and in 1691 he published a book entitled *Curiosities in Chymistry: Being New Experiments and Observations Concerning the Principles of Natural Bodies*, which repeatedly invoked Boyle's authority and which the title page even attempted to father on Boyle himself through the formula: 'Written by a Person of Honour, and Published by his Operator,

---

[34]   See R.E.W. Maddison, 'Boyle's Operator: Ambrose Godfrey Hanckwitz, F.R.S.', *NRRS*, 11 (1955), 159–88; Principe, *Aspiring Adept*, pp. 134–6.

[35]   See *Boyle Papers*, pp. 50–51; Hillel Schwartz, *The French Prophets: The History of a Millenarian Group in Eighteenth-century England* (Berkeley, CA and Los Angeles, CA, 1980), esp. ch. 4.

[36]   H.J. Cook, 'Sir John Colbatch and Augustan Medicine: Experimentalism, Character and Entrepreneurialism', *Annals of Science*, 47 (1990), 475–505, on p. 496n.; Thomas Emes, *A Dialogue between Alkali and Acid* (London, 1698), pp. 65–6 and passim; Thomas Emes, *A Letter to a Gentleman concerning Alkali and Acid* (London, 1700); Schwartz, *French Prophets*, p. 250.

[37]   Robert St Clair, *The Abyssinian Philosophy Confuted* (London, 1697), sig. a4 and 'To the Reader', passim. The body of the book takes the form of a translation of a work by Bernadino Ramazzini. Along with Greg, Slare and others, St Clair is referred to in Boyle's will: R.E.W. Maddison, *The Life of the Hon. Robert Boyle, F.R.S.* (London, 1969), p. 262.

H.G.'. On the other hand, the book comprises a curious mixture of ideas, some Boylean and some derived from contemporary anatomical writings, whereas others reflect the influence of Paracelsian and other theories, including J.B. van Helmont's view of water as the ultimate principle of nature which Boyle had tested and found wanting in a classic passage in *The Sceptical Chymist* (1661).[38] Indeed, though Boyle himself may have been notable for his eclecticism, this was by no means unique to him – even if none of these lesser men achieved as influential a synthesis as his by such means.

Yet this raises the broader issue of how Boyle was received by his contemporaries, another topic that cries out for fuller attention than hitherto, and a few remarks may be made on this subject here. In contrast to most of his intellectual associates, Boyle lacked a university education and, though he did much to make up for this, he was always something of an outsider to academic life. Thus he felt it necessary to make clear that he was 'a person that never was a Gown-man', and it is perhaps in this connection that one should see the criticisms that were made at the time of defects in his style and vocabulary, not least by his Oxford colleague John Wallis in a document sent to Boyle in 1663.[39] Interestingly, a similar remark was made by the naturalist John Ray in a letter to the virtuoso John Aubrey in 1691, in which he took issue with certain words that Aubrey used in his *Natural History of Wiltshire*, adding: 'You are not ignorant how Mr Boyl hath been κωμωδούμενος [a subject of ridicule] for some new-coyn'd words such as ignore & opine.'[40] It is revealing that Aubrey – himself something of an intellectual outsider despite his Oxford education – was paired with Boyle in this way, and, though we may enjoy the rich crop of neologisms that Boyle used in his writings, it is important to note that not all at the time were at ease with this, thus illustrating a complication in his relations with his context which it would be easy to overlook.[41]

---

[38] Greg, *Curiosities in Chymistry* (London, 1691). For a commentary, see Bruce T. Moran, *Distilling Knowledge: Alchemy, Chemistry and the Scientific Revolution* (Cambridge, MA, 2005), pp. 94–8. On the 'willow tree' experiment, see Charles Webster, 'Water as the Ultimate Principle of Nature: The Background to Boyle's *Sceptical Chymist*', *Ambix*, 13 (1966), 96–107.

[39] *Works*, vol. 7, p. 141, *Correspondence*, vol. 2, pp. 112ff., and see *Scrupulosity and Science*, p. 137, *Between God and Science*, p. 142.

[40] R.W.T. Gunther (ed.), *Further Correspondence of John Ray* (London, 1928), p. 170. In Ray's letter to Aubrey of 27 October 1691, included in the MS of the *Natural History of Wiltshire*, Bodleian Library, Oxford, MS Aubrey 1, fol. 13 (printed in John Britton's 1847 edition of the work, p. 8), Ray says of the unusual words that Aubrey used: 'I know not whether the wits will allow them.'

[41] On Boyle's neologisms, see Maurizio Gotti, *Robert Boyle and the Language of Science* (Milan, 1996), ch. 3. For a more positive view of Boyle's literary qualities, see Sir Peter Pett's comments in *Boyle by Himself and His Friends*, pp. 69–70.

It is also worth bearing in mind that during his lifetime, Boyle was most widely read not as a natural philosopher, but as a religious writer. By far his most popular book was his devotional work *Some Motives and Incentives to the Love of God*, familiarly known as *Seraphic Love*, which had originally been compiled during his moralist period in the 1640s but was greatly extended, often with rather disparate material, prior to its publication in 1659.[42] Between then and 1708, the book ran to 10 editions in English (one of them issued by four separate publishers), and it was also translated into French in 1671, German in 1682 and Latin in 1693.[43] There would thus be a case for seeing Boyle as a lay religious writer comparable to an aristocrat like Anne Douglas, Countess of Morton, whose *Daily Exercise* (1665) has been included among the 'religious best sellers' of Restoration England, and this was a role that Boyle was happy to adopt.[44] Indeed, though most such best sellers were in fact written by churchmen, Boyle specifically noted concerning religious books how 'it has been observ'd, that those penn'd by Lay-men, and especially Gentlemen, have (*cæteris paribus*) been better entertain'd, and more effectual than those of Ecclesiasticks'.[45]

By comparison, most of Boyle's scientific writings only ever ran to one edition in English, though a handful achieved two and just two such works ran to three – namely his first scientific book, *New Experiments Physico-Mechanical, Touching the Spring of the Air and its Effects* (1660) and the first 'tome' of *Some Considerations touching the Usefulness of Natural Philosophy* (1663).[46] In fact, Boyle's most reprinted work after *Seraphic Love* was his collection of recipes, posthumously published as *Medicinal Experiments*, which (ignoring Boyle's private trial edition of it in 1688) ran to seven editions between 1692 and 1731.[47] On the other hand, the relative popularity of *Medical Experiments* and *Usefulness*, and not least the latter, is itself revealing, since, ironically, they seem to have appealed widely for reasons with which Boyle was slightly ill at ease. For all its significance as a manifesto for the potential utility of scientific findings

---

[42]   See Lawrence M. Principe, 'Style and Thought of the Early Boyle: Discovery of the 1648 Manuscript of *Seraphic Love*', *Isis*, 85 (1994), 247–60.

[43]   See J.F. Fulton, *A Bibliography of the Hon. Robert Boyle, F.R.S.* (2nd edn, Oxford, 1961), pp. 2–9. Of Boyle's other religious writings, *Style of the Scriptures* ran to four English editions: ibid., pp. 32–5.

[44]   C. John Sommerville, *Popular Religion in Restoration England* (Gainesville, FL, 1977), p. 39.

[45]   Ibid., ch. 3, passim; *Works*, vol. 8, p. 8.

[46]   See Fulton, *Bibliography*, passim. Strictly speaking, the different versions of tome 1 of *Usefulness* are issues rather than editions, but there were three separate settings of type, one almost certainly produced in conjunction with the publication of the second tome in 1671: ibid., pp. 37–41.

[47]   Ibid., pp. 118–24. For the 1688 edition, see below, pp. 156–7, and *Works*, vol. 11, pp. xxvii–xxix, 173ff.

to all aspects of human life, much of the content of *Usefulness* comprises rather miscellaneous, often anecdotal, material, including a vast number of recipes. It is therefore not surprising that Boyle felt the need to apologise at one point in the book that 'it may appear somewhat below me, in a Book, whose Title seems to promise you Philosophical Matters, to insert I know not how many Receipts'.[48] Similarly, in connection with *Medicinal Experiments*, he explained how in publishing it he had to overcome 'the Reluctancy which a natural tendernes for reputation had given me to suffer things that many will looke upon as tryfles to appear publickly in a criticall age'.[49] Yet it is clear that it was precisely these aspects of the works in question that were widely popular with his contemporaries.

A case in point is provided by the commonplace book of Edmund Kershaw, a member of a Lancashire family of middling status, which is now in Princeton University Library.[50] This contains a variety of material, including legal documents and templates and astrological charts complete with moving dials and pointers. It also contains extensive notes from one of the classic books of secrets that circulated in seventeenth-century England, Dr Richard Read's English version of Johann Jacob Wecker's *De secretis libri XVII* (1582), published in 1660.[51] But alongside these are many lengthy extracts from Boyle's *Usefulness of Natural Philosophy*, and what is striking is that these are presented as if they came from a similar book of secrets, replete with titles like 'Jaundice Yellow how to be cured' or 'How to preserve fish from Stinking in Spirit of Wine' or 'The Cure of a Person esteemed Bewitched by an appended Mineral'.[52] It is true that Boyle himself encouraged this to some extent, since a few of these titles derive from his text; on the other hand, most are extraneous, and it is also interesting that, in instances where Boyle included ancillary remarks that distanced him from the 'superstitious' aspects of a remedy he cited, Kershaw often ignored these.[53] What is worthy of note is how easily a reader like Kershaw could adopt

---

[48] Ibid., vol. 3, p. 476.

[49] RS MS 186, fols 120v–121, quoted in *Scrupulosity and Science*, p. 215.

[50] Princeton University Library MS CO199, no. 553. For the Kershaw family, see *The Victoria County History of Lancashire*, vol. 6 (1911), p. 167, and W.A. Abram, 'Ancient Lancashire Families. IV – The Kershaws of Rochdale and of Heskin Hall in Eccleston', *Lancashire and Cheshire Antiquarian Notes*, 2 (1886), 132–40.

[51] R. Read (trans.), *Eighteen Books of the Secrets of Art & Nature* (London, 1660). For the history of this work, see William Eamon, *Science and the Secrets of Nature* (Princeton, NJ, 1994), esp. pp. 276–8. For Kershaw's extracts from it, see CO199, no. 553, passim. It seems likely that other extracts come from a comparable book or books yet to be identified.

[52] CO 199, no. 553, fols 106 (cf. *Works*, vol. 3, pp. 430–31), 109 (cf. ibid., p. 310), 112v (cf. ibid., p. 422).

[53] For the former, cf. CO 199, no. 553, fols 108, 110, 111, with *Works*, vol. 3, pp. 484–5, 489–91. For the latter, cf. CO 199, no. 553, fols 112, 119v, 'Witchcraft how Remedied & cured', with *Works*, vol. 3, pp. 423–4.

Boyle into a context to which we might see him as alien and with which his own remarks suggest that he was somewhat ill at ease.

A perhaps more predictable reaction to Boyle is to be found in a section entitled 'A Miscellany of Experiments. Liber the I' in another commonplace book, this time compiled in 1690 by George Fleming, a member of a distinguished Westmoreland family who was then an undergraduate at Oxford; later, he was to go on to become successively Dean and Bishop of Carlisle in the early eighteenth century.[54] We know from accounts preserved among the rich archival records of his family that Fleming owned a number of Boyle's books, including *The Sceptical Chymist* and *Of the Reconcileableness of Specifick Medicines to the Corpuscular Philosophy* (1685).[55] The 54 numbered extracts from Boyle with which his 'Miscellany of Experiments' begins are taken from three works: *Certain Physiological Essays* (1661), *Cosmical Qualities* (1670) and, as with Kershaw, *The Usefulness of Natural Philosophy*. It is revealing that in practice Fleming's extracts comprise not so much the careful experiments for which Boyle is famous – as his title implies – but rather recipes, observations about the properties of things and the like, bearing some resemblance to the kind of material that interested Kershaw. Indeed, it may be that it was in the nature of compilations of this kind to encourage such miscellaneity, since this is equally true of Fleming's accompanying extracts from *Philosophical Transactions*, from Robert Plot's *Natural History of Oxfordshire* (1677) and from Nehemiah Grew's *Musæum Regalis Societatis* (1681). But what is revealing in Fleming's case is how his title picked up the crucial concept of 'Experiment', and it could be argued that what Boyle seems to have encouraged in this instance was an omnivorous curiosity of the kind that is reflected in so many of his own books and which is clearly to be seen as a crucial part of his impact on the science of his day.

Indeed, one of the things that was notable about Boyle was the sheer breadth of his interest in natural phenomena, and it is not surprising that this should be a key part of his influence. His publications covered an extraordinary range of topics – from the nature of colours or cold to that of the air, and from hydrostatics or the structure of crystals to the workings of the human body. His workdiaries were no less broad in their remit: in addition to showing him carrying out controlled experiments on a wide range of phenomena, they are equally full of telling information divulged to him by others. As we will see in Chapter 9, one important component of this comprises reports from travellers that Boyle intended to redeploy in 'Physica Peregrinans', covering everything from atmospheric conditions to strange animals. For all his emphasis on the

---

[54] Kendal Archive Centre, WDRY/3/1/1; J.R. McGrath (ed.), *The Flemings at Oxford* (3 vols, Oxford Historical Society, vols 44 (1903), 62 (1913) and 79 (1924)), vol. 2, pp. vi–viii and passim; vol. 3, pp. ix–x and passim; *ODNB*.

[55] McGrath, *Flemings at Oxford*, vol. 2, pp. 126, 290, 298; vol. 3, pp. 26, 35.

crucial importance of experiments, Boyle laid an equal stress on 'observations' as the factual, experiential basis from which scientific conclusions should be derived, constantly emphasising 'particulars' as the necessary corrective to facile generalisation.[56] It seems likely that many contemporaries were receptive to this and valued his writings accordingly.

If some responded to Boyle at this rather basic level, however, it is clear that others engaged more fully with the agenda for which he is retrospectively remembered – particularly his vindication of carefully controlled and fully recorded experiments, and his role as one of the leading protagonists of the mechanical philosophy. Here, a valuable case study has recently been provided by Sue Hemmens in her account of the chemical agenda of the Dublin Philosophical Society in its early years, since it turns out that this was strongly influenced by Boyle, in terms both of his basic postulate of the particulate nature of matter and of his use of systematic experiment, which its members applied to the analysis of mineral waters and of blood and used as the basis of a Boylean critique of the acid-alkali hypothesis.[57] It is also clear that Boyle had a significant impact on the chemical investigations carried out in the later years of the seventeenth century and the early years of the eighteenth by figures like Samuel Cottereaux du Clos and Wilhelm Homberg at the Académie des Sciences in Paris and Herman Boerhaave at the University of Leiden.[58] In the case of du Clos, as we will see in Chapter 6, there is evidence that he was rather irritated by some of the traits in Boyle that evidently commended him to readers like Kershaw, such as his arch recourse to secrecy when it suited him, and this implies a certain tension between the clienteles to which Boyle appealed which is suggestive in itself.

Overall, Boyle's influence among readers throughout Europe cries out for fuller exploration than it has yet received: it was precisely to ensure that his natural philosophical works were as widely accessible as possible that Boyle took such trouble to have his books translated into Latin throughout his career, as is explained in Chapter 7, and there is reason to believe that this concern on his part paid off to a significant extent. His profound writings about the philosophical

---

[56] For a helpful discussion of Boyle's view of the respective role of experiments and observations, see Rose-Mary Sargent, *The Diffident Naturalist: Robert Boyle and the Philosophy of Experiment* (Chicago, IL, 1995), ch. 6, esp. pp. 136–7.

[57] Sue Hemmens, 'Crow's Nest and Beyond: Chymistry in the Dublin Philosophical Society, 1683–1709', *Intellectual History Review*, 25 (2015), 59–80. See also below, p. 105.

[58] See Lawrence M. Principe, 'Liens et Influences Chimiques entre Robert Boyle et la France', in Myriam Dennehy and Charles Ramond (eds), *La Philosophie Naturelle de Robert Boyle* (Paris, 2009), pp. 71–89; Rémi Franckowiak, 'Du Clos, un Chimiste *post-Sceptical Chymist*', in ibid., pp. 361–77; Victor D. Boantza, 'Chymical Philosophy and Boyle's Incongruous Philosophical Chymistry', in Ofer Gal and Raz Chen-Morris (eds), *Science in the Age of Baroque* (Dordrecht, 2012), pp. 257–84, and John C. Powers, *Inventing Chemistry: Herman Boerhaave and the Reform of the Chemical Arts* (Chicago, IL, 2012), passim.

aspects of science and about the relationship between God and nature also had a legacy, both in a European context and in an English one. For instance, the impact of his *Free Enquiry into the Vulgarly Received Notion of Nature* (1686) on both Gottfried Wilhelm Leibniz and Johann Sturm has recently been reappraised, while a study of his writings on things above reason has stressed the extent to which it was Boyle's formulation that was the target of attacks by Deist writers like John Toland and Anthony Collins in the generation after his death.[59] Undoubtedly, there is much still to be learned about Boyle's powerful influence on the intellectual developments of his age, and it seems likely that this will only increase our estimate of his significance for the evolution of ideas in the early years of the Enlightenment.

Postscript: Genius Eclipsed

Yet there was a definite decline in Boyle's reputation as the eighteenth century progressed, and it is appropriate to comment on this by way of an epilogue to this introduction.[60] Of course there was an extent to which Boyle's reputation remained high, particularly in the religious sphere, where his legendary piety retained its exemplary value. Indeed, his iconic role became all the more crucial as anxieties were expressed about the religious credentials of certain scientific practitioners, and even in the early nineteenth century he continued to be promoted in precisely this connection.[61] He also had a continuing influence in more specific ways, for instance in the genre of 'meletetics' that he had exemplified in his *Occasional Reflections upon Several Subjects* (1665) – another of his religious writings which was well received even if it failed to achieve the publishing success of *Seraphic Love*, running to only two editions. As Robert Mayhew has pointed out, the 'procedure of meditative perception of nature' that Boyle exemplified remained popular right through the eighteenth century in works like James Hervey's *Meditations and Contemplations* (1746–47) and even

---

[59] See Myriam Dennehy, 'Leibniz et Sturm, Lecteurs de Boyle', in Dennehy and Ramond, *La Philosophie Naturelle de Robert Boyle*, pp. 331–59; Thomas Holden, 'Robert Boyle on Things above Reason', *British Journal for the History of Philosophy*, 15 (2007), 283–312.

[60] For an important recent study which gives greater precision to the overall picture presented here, see Peter Anstey, 'Experimental Pedagogy and the Eclipse of Robert Boyle in England', *Intellectual History Review*, 25 (2015), 115–31.

[61] See, for instance, Henry Rogers' 'Introductory Essay' to his edition of *Boyle's Treatises* [*High Veneration, Things above Reason* and *Style of the Scriptures*] (London, 1835), esp. pp. xlvff. See further *Scrupulosity and Science*, pp. 262–3, and see also J.R. Wigelsworth, *Deism in Enlightenment England: Theology, Politics and Newtonian Public Science* (Manchester, 2009), p. 193.

in the writings of the great advocate of the picturesque William Gilpin.[62] Boyle's influence also continued in more miscellaneous areas, for instance in terms of the use of questionnaires in travel literature, as briefly surveyed in Chapter 3.

As a natural philosopher, however, Boyle's reputation soon began to be eclipsed, as I illustrated in *Boyle: Between God and Science* by citing the contrast between the 'Hermitage', a structure erected in the royal gardens at Richmond by Queen Caroline, wife of George II, between 1731 and 1733 – where a bust of Boyle received pride of place – and the Temple of British Worthies at Stowe in Buckinghamshire, a comparable architectural composition erected shortly afterwards, in which Boyle failed to appear at all.[63] At the Hermitage, Boyle's bust was accompanied by those of Newton, Locke, Newton's protégé Samuel Clarke and the religious writer William Wollaston, but his 'stands higher than these, on a Pedestal, in the inmost, and, as it were, the most sacred Recess of the Place; behind his Head a large Golden Sun, darting his wide spreading Beams all about, and towards the others, to whom his Aspect is directed'.[64] It was literally Boyle's apotheosis – the way he was placed in an apse with a sunburst behind him was reminiscent of Bernini's 'Ecstasy of St Teresa' – and no less striking was the arrangement of the ensemble to give him pre-eminence as a prolific and innovative scientist who at the same time illustrated the harmony between science and religion.

But the adulation represented by the Hermitage was not to be repeated. Quite apart from Boyle's absence from the Temple of British Worthies at Stowe, it is also symptomatic that when busts of various modern celebrities were commissioned a few years later for the library at Trinity College, Cambridge, Boyle was not among them. Instead, in both cases the role of leading scientist was taken by Newton, and Newton has dominated perceptions of the science of the period ever since. The result was the increasing marginalisation of Boyle in eighteenth-century English culture, and the reasons for this are worth exploring.

The problem had started when Newton began to be acclaimed as a scientific genius, particularly for the comprehensive solution to the problems of celestial mechanics represented by his *Principia* (1687). This book was already described by John Aubrey before his death in 1697 as 'the greatest highth of Knowledge that humane nature has yet arrived to', an appraisal probably reflecting a consensus view that Aubrey had picked up, notwithstanding his own mathematical interests.[65] Ironically, such views were reinforced by the institution founded under the provisions of Boyle's will, the Boyle Lectures, and

---

[62] R.J. Mayhew, *Landscape, Literature and English Religious Culture, 1660–1800: Samuel Johnson and the Language of Natural Description* (Basingstoke, 2004), ch. 4.

[63] See *Between God and Science*, pp. 251–2, and the references given in ibid. pp. 348–9.

[64] *Gentleman's Magazine*, 3 (1733), 208.

[65] See Michael Hunter, *John Aubrey and the Realm of Learning* (London, 1975), p. 60. For Aubrey's mathematical studies, see ibid., esp. pp. 48–50

particularly the first series delivered in 1692 by the scholar and divine Richard Bentley. Though Bentley did not neglect Boyle, his lectures have rightly been seen as significant for the space they devoted to an exposition of the findings of 'that very excellent and divine Theorist Mr. *Isaac Newton*, to whose most admirable sagacity and industry we shall frequently be obliged in this and the following Discourse'.[66] Newton's pre-eminence was further intensified by the acclaim for his experimental findings concerning light and related phenomena in his equally seminal *Opticks* when this came out in 1704. Indeed, in a sense this was more of a threat to Boyle than the *Principia*, since, in contrast to the mathematical abstrusities of Newton's first book, this presented Newton as a champion of hands-on experiment, which had the effect of rendering a key facet of Boyle's role superfluous.

The process by which Newton's scientific reputation grew as Boyle's declined was exacerbated by the extent to which it became modish to invoke Newton on natural philosophical issues in the early eighteenth century, something which even Boyle's supporters unwittingly abetted. Thus, in his perceptive life of Boyle, mainly written in the first decade of the early eighteenth century but unfortunately never completed, the scholar William Wotton explained how Boyle would have made better sense of certain of his findings with his air-pump had he been aware of the Newtonian concept of attraction: 'But the Discovery of that general Law of Matter was not then known, that was reserved to the incomparable Sir Isaac Newton to find out.'[67] Similarly, in the most important version of Boyle's writings to be published at this time, the epitome of them produced in 1725 by the scientific writer and lecturer Peter Shaw, it was repeatedly emphasised how Boyle's discoveries had been furthered and enhanced by Newton, 'whose words we can never use too much'.[68]

To some extent there may have been an element of political correctness in this: undue praise for Boyle might have been seen as an implicit criticism of his successor, and it is certainly the case that Newton's chief Continental rival, G.W. Leibniz, championed what he claimed to be Boyle's version of the mechanical philosophy against Newton's theory of matter, which cannot have helped Boyle in the fervently Newtonian atmosphere of Augustan England.[69] Probably more important, however, was the extent to which this was the result of the innate presentism to which science was as prone then as it has been ever

---

[66]   Richard Bentley, *A Confutation of Atheism from the Origin and Frame of the World. Part II* (London, 1693), in I.B. Cohen (ed.), *Sir Isaac Newton's Papers and Letters on Natural Philosophy* (2nd edn, Cambridge, MA, 1978), p. 320.

[67]   *Boyle by Himself and His Friends*, p. 145.

[68]   Peter Shaw (ed.), *The Philosophical Works of the Hon. Robert Boyle* (3 vols, London, 1725), vol. 1, p. 386, quoted in *Scrupulosity and Science*, p. 262.

[69]   See Antonio Clericuzio, 'A Redefinition of Boyle's Chemistry and Corpuscular Philosophy', *Annals of Science*, 47 (1990), 561–89, on pp. 561–2.

since – of Boyle simply seeming out of date due to the recent developments in science for which Newton and his followers were responsible. This was well put by one commentator, John Weyland, in 1808, when he explained how, for all 'the *acknowledged* importance of Mr Boyle's physical discoveries; considering the state of science at the time he lived' the value of findings like his had been 'extremely reduced' by 'the modern discoveries'.[70]

In the case of Boyle and Newton, however, it might be argued that more was at stake than this, and that Boyle was the victim of a crucial shift in expectations concerning the proper nature of science, with a new paradigm emerging in the image of Newton which discredited the entire style of science that had preceded him and that Boyle had so fully exemplified. Boyle's writings are voluminous, prolix and disorganised. He was even reluctant to countenance a collected edition of them during his lifetime, though this situation was rectified by epitomes like Shaw's and subsequently by the multi-volume edition produced by Thomas Birch in 1744 – even if it seems likely that copies of the latter were purchased more as imposing library furniture than to be read. It was partly in order to rectify the often rather diffuse and miscellaneous character of Boyle's writings that in his version Shaw attempted to 'methodise' Boyle's work, demonstrating how it could be seen as systematic in its coverage of natural phenomena and thus appear properly 'philosophical' in tone.[71]

By contrast, Newton's *Opticks* – the work by Boyle's successor that was most likely to be compared with his – has always rightly been acclaimed for its brilliant combination of extraordinary range with great succinctness. Insofar as it included broader and more speculative ideas, these appeared as a concluding series of 'Queries' (extended in successive editions), whereas the core of the work was set out in quasi-mathematical style as a series of 'propositions', preceded by even more analytical 'definitions' and 'axioms'. Such a structure was of course still more overt in the *Principia*, where it was accompanied by a formidable mathematical complexity to which Boyle was largely a stranger. (For all his fascination with quantification as a scientific tool, Boyle was distrustful of the higher forms of mathematics, which he thought involved a degree of idealisation which made them remote from reality.[72]) In *Opticks*, however, Newton made such a mode of presentation equally appropriate in relation to experimental

---

[70] John Weyland (ed.), *The Hon. Robert Boyle's 'Occasional Reflections'. With a Preface &c.* (London, 1808), p. viin., quoted in *Scrupulosity and Science*, p. 261

[71] See Harriet Knight, 'Rearranging Seventeenth-century Natural History into Natural Philosophy: Eighteenth-century editions of Boyle's Works', in David Knight and Matthew Eddy (eds), *Science and Belief: From Natural Philosophy to Natural Science, 1700–1900* (Aldershot, 2005), pp. 31–42.

[72] See Steven Shapin, 'Robert Boyle and Mathematics: Reality, Representation and Experimental Practice', *Science in Context*, 2 (1988), 23–58.

data, and this stood in stark contrast to the often rather discursive character of
the treatises of Boyle.

There was also a contrast in intellectual style, and this is true not only of
Boyle, but also of his contemporaries like Robert Hooke. As we have already
seen, Boyle was fascinated by almost everything and Hooke displayed a similarly
wide curiosity, on topics ranging from comets to the art of navigation, from
the nature of light to petrifaction and fossils. This is well exemplified by his
most famous book, *Micrographia* (1665), which, quite apart from the stunning
exposition of the findings of the microscope for which it is famous, also included
speculations on all sorts of topics, even including the surface of the Moon.
Indeed, to a contemporary of Hooke's like John Aubrey, this sheer fecundity
was one of the chief marks of his genius.[73] By comparison, Newton was more
single-minded, concentrating on the solution of great but finite problems – of
celestial mechanics or of the nature of light – and the implication was to make
his predecessors seem like dilettantes, incapable of the profound insights which
rewarded the intensity of the scrutiny in which he engaged.

There is a further corollary to this in terms of overall philosophical outlook,
in that such single-mindedness on Newton's part was combined with a perhaps
slightly arrogant conviction that God had vouchsafed to him and a select few
an ability to penetrate the secrets of the universe which lesser mortals lacked.
Boyle, on the other hand, preached a general humility concerning the ability
of mankind as a whole to penetrate the full profundity of God's design in the
universe. This difference, too, arguably had an indirect influence on the scientific
paradigm that Newton bequeathed to posterity.[74]

In such contrasts there is inevitably an element of caricature. In fact, a good
deal of eighteenth-century natural philosophy remained quite Boylean in spirit,
and some of Boyle's speculations on a wide range of topics – from the temperature
of the submarine regions to the effect of the environment on human health – had
an influence that lasted perhaps surprisingly long.[75] But there can be no question
that it was not just the modish championship of Newtonian ideas that had the
effect of encouraging Boyle's memory to fade in the eighteenth century, but also
the triumph of a new image of science in the Newtonian mould – a vision of the
relentless pursuit of truth by a remote and single-minded genius, resulting in
the establishment of overriding 'principles' which encapsulated the general laws
of nature.

---

[73]   See John Aubrey, *Brief Lives, Chiefly of Contemporaries*, ed. Andrew Clark (2 vols,
Oxford, 1898), vol. 1, pp. 409–16.

[74]   It has not seemed appropriate to try to summarise here the immense literature
relating to Newton which this section takes for granted. For a recent introduction, see Denis
R. Alexander (ed.), *The Isaac Newton Guide Book* (Cambridge, 2012).

[75]   See *Between God and Science*, pp. 173–4, and the references in ibid., pp. 333–4.

Our task in relation to Boyle, on the other hand, is to restore prominence to those aspects of his science which the Newtonian paradigm tended to occlude – his endless curiosity about how the world worked and how it might be improved, and his pursuit of a mass of particulars as the basis of a true understanding of nature. It is also right to do justice to the diffidence about explanations that characterised Boyle and his like and the serendipity of experiment and observation that characterised such men, and to emphasise the heroic open-mindedness that formed part of their crusade against the dogmatism of the scholastic paradigm which, ironically, Newtonianism was prone in some respects to revive.[76] In addition, however, as the subsequent chapters in this book demonstrate, it is crucial to achieve detailed knowledge of a thinker like Boyle in all his complexity – giving proper attention not only to his intellectual ambitions, but also to the complications in his ideas, the influences to which he was subject and the constraints by which he was affected. Only by this means will we come to a true understanding of his seminal role at a transitional moment in Western intellectual history.

---

[76] A case study is provided in Michael Hunter, 'Boyle's Legacy: Second Sight in English and Scottish Thought in the long Eighteenth century', delivered at the Second Sight and Prophecy Conference, University of Aberdeen, 14–16 June 2013.

Appendix

Boyle's Desiderata for Science

Boyle's 'wishlist' for science has attracted a good deal of public attention since it was included in the Royal Society's summer science exhibition in 2010, even featuring in Brian Cox's 2013 television series *Science Britannica*.[1] Interest has focused particularly on the question of which of Boyle's hopes for outcomes from science have since been realised – in all but two of his desiderata, in Professor Cox's view – and Boyle's prescience has received appropriate acclaim, rather as has traditionally been the case with the vision of science's potential displayed by his mentor, Francis Bacon, in his *New Atlantis*.[2] In view of the degree of interest the document has aroused, it seems appropriate to print the fullest version of it here (not in fact the version that has attracted public attention in recent years, but a later one), annotated with such information as is available to throw light on what Boyle might have been referring to in his various suggestions.

By way of background, it should be pointed out that this document has a context in terms of desiderata lists for science going back to Bacon that has recently been sketched by Vera Keller.[3] In Boyle's case, lists of memoranda form a significant component of his manuscript remains that deserve more attention than they have yet received: these are rather different both from the sets of 'heads' and 'inquiries' that I have published elsewhere and from the outlines of putative books that are also to be found within the Boyle archive.[4] A further analogue for the second part of the document is provided by the 'Designe about Natural History' that Boyle compiled in the 1660s: among the miscellaneous papers relating to this is one that sets out a programme for the promotion of science which bears some resemblance to the prescriptions made here, though

---

[1]   See, for example, Ian Sample, 'Robert Boyle: Wishlist of a Restoration Visionary', *The Guardian*, 3 June 2010; Christine Patterson, 'Why We are Shamed by Robert Boyle's Pursuit of Knowledge', *The Independent*, 5 June 2010; Brian Cox, 'Clear Blue Skies', programme 3 of *Science Britannica*, broadcast on BBC2 on 2 October 2013.

[2]   See, for example, W.P.D. Wightman, *Science in a Renaissance Society* (London, 1972), p. 161, who observes that Bacon's work 'reads like a prospectus for ... the Low Temperature Research Association (founded at Cambridge) and the Scottish Hydroelectric Board'.

[3]   See Vera Keller, 'The "New World of Sciences": The Temporality of the Research Agenda and the Unending Ambitions of Science', *Isis*, 103 (2012), 727–34.

[4]   See Michael Hunter (ed.), *Robert Boyle's 'Heads' and 'Inquiries'*, Robert Boyle Project Occasional Papers No. 1 (London, 2005), p. viii and passim.

these are more elaborate either than that or than any proposals that Boyle made elsewhere.[5]

Even more relevant is a list of 'Optatives' in the second 'tome' of Boyle's *Usefulness of Natural Philosophy* (1671) which deals with desiderata ranging from making iron fusible to the use of moulded wood, and which bears some similarity to the current document. Indeed, Boyle's commentary on the *Usefulness* list is almost equally applicable to this one, and it seems worth quoting it here: like many passages in *Usefulness*, it is addressed to 'Pyrophilus', in other words, Boyle's nephew, Richard Jones, later Earl of Ranelagh:

> It now remains, that I mention in a few words the Optatives, that may be propos'd by the Naturalist about the particular Trades he would improve. By which name of Optatives, I mean all those Perfections, that being desirable, are rather very difficult, than absolutely impossible, to be obtain'd ....
>
> I know, *Pyrophilus*, that such Optatives may be thought but a civill name for Chymerical Projects; but I shall hereafter more fully declare to you, why I think it not altogether unuseful, that such Optatives should be propos'd, provided, as I hinted above, that they be very difficult, & not impossible: That is, that they be such, as are not repugnant to the nature of the things, nor the general Principles of Reason and Philosophie, and seem no otherwise to be Chymically or Mechanically impossible, than because we want Tooles or other Instruments and wayes to perform some things necessary to the compassing of the propos'd End, or to remove some difficulties, or remedy some Inconveniences, that are incident to us in the Prosecution of such difficult designs.
>
> And let me here tell you, *Pyrophilus*, that this Advantage may be deriv'd from the deviseing of such Optatives to bold and sagacious Men, that if they despair of attaining to the Perfection they are invited to aime at, they may at least endeavour to reach some Approximation to it.[6]

More specific information relating to each of Boyle's desiderata is given in the notes that follow, but one last general point may be made here, namely that, though there is no doubt that the document is properly associated with Boyle, it is striking how often its elucidation leads towards his protégé and colleague Robert Hooke, and in some cases towards their close associate both at Oxford

---

[5] See Michael Hunter and Peter Anstey (eds), *The Text of Robert Boyle's 'Designe about Natural History'*, Robert Boyle Project Occasional Papers No. 3 (London, 2008), pp. 11–12 and passim.

[6] *Works*, vol. 6, pp. 480–81. I am grateful to Ted Davis for reminding me of this passage. For the concept of 'optatives', see Vera Keller, 'Accounting for Invention: Guido Pancirolli's Lost and Found Things and the Development of *Desiderata*', *Journal of the History of Ideas*, 73 (2012), 223–45, esp. pp. 237–8.

and London, Christopher Wren – thus representing a tribute to Boyle's interest in scientific developments in which he was not directly involved himself.[7]

The text presented here is taken from Boyle Papers 36, fols 77v–78. It is a fair copy in the hand of Robin Bacon, the amanuensis Boyle most frequently used in his later years, laid out on the facing inner pages of a bifoliate, the outer sides of both of which are blank. The text of the first section of the list, on fol. 77v, has been collated with the version of this that survives in BP 8, fols 208–9 (as displayed in the 2010 exhibition). That version, which lacks the text that appears on fol. 78 in BP 36, is in hand H, dating from the 1660s;[8] the only differences between the two versions are trivial ones of orthography and capitalisation that have here been ignored.

The Prolongation of Life.[9]
The Recovery of Youth, or at least some of the Marks of it, as new Teeth, new Hair colour'd as in Youth.[10]
The Art of flying.[11]
The Art of continuing long under water, and exercising functions freely there.[12]

---

[7] For a parallel, see his interest in astronomical observations as discussed in Chapter 4 below. For a comparable list of desiderata compiled by Wren, see Christopher Wren jun. (ed.), *Parentalia: or, Memoirs of the Family of the Wrens* (London, 1750), pp. 198–9.

[8] See *Boyle Papers*, p. 55. For Bacon, see ibid., pp. 47–8.

[9] This is perhaps the classic ambition of science, shared by Boyle with Bacon and many others. Indeed, this and the next phrase echo the opening of Bacon's list of *Magnalia naturæ* at the end of *New Atlantis*. For a general account, see David B. Haycock, *Mortal Coil: A Short History of Living Longer* (New Haven, CT and London, 2008), pp. 35, 60–68 (on Boyle) and passim.

[10] Boyle collected various recipes for preserving teeth, as expounded in his *Medicinal Experiments*, but is not known to have investigated their renewal. As for hair, he was aware that its colour could suddenly change (see, for example, *Works*, vol. 3, p. 444; vol. 10, pp. 316–17), but there is no evidence that he sought to manipulate this. However, for a discussion of rejuvenescence in a Paracelsian context, in which both teeth and hair are referred to, see *Works*, vol. 3, pp. 408–9.

[11] I have been unable to discover any reference to human flight on Boyle's part, but for Robert Hooke's interest in it, see Michael Hunter, 'Hooke the Natural Philosopher', in Jim Bennett et al., *London's Leonardo: The Life and Work of Robert Hooke* (Oxford, 2003), pp. 105–62, on pp. 149–50. I have also benefited from reading an unpublished paper on this topic by Nicole Howard.

[12] Boyle repeatedly investigated the experiences of divers, and not least the extent to which they were affected by underwater conditions. See esp. *Works*, vol. 5, pp. 270–77; vol. 6, pp. 347ff., 482; vol. 7, pp. 168–83, 220–21. For an overview from the point of view of testimony, see Steven Shapin, *A Social History of Truth* (Chicago, IL, 1994), pp. 258–66. Boyle's interest in such matters is reflected by many reports in Workdiaries 21 and 36, while

The Cure of Wounds at a Distance.[13]
The Cure of Diseases at a Distance, or at least by Transplantation.[14]
The attaining Gigantick Dimensions.[15]
The Emulating of Fish without Engines by Custome and Education only.[16]
The Acceleration of the production of things out of seed.[17]
The Transmutation of Metals.[18]
The making of Glass Malleable.[19]
The Transmutation of Species in Mineralls, Animals, and Vegetables.[20]
The Liquor Alcahest and other dissolving Menstruums.[21]
The making of Parabolicall & Hyperbolicall glasses.[22]
The making of Armour Light, and extreamly hard.[23]

---

also relevant are his comments on the submarine of Cornelius Drebell: see *Works*, vol. 1, pp. 287–8; vol. 6, pp. 481–2.

[13] Presumably an allusion to such cures as expounded by J.B. van Helmont and by Boyle's one-time associate Sir Kenelm Digby, though Boyle was somewhat sceptical: see *Works*, vol. 3, p. 410. For recent accounts, see Elizabeth Hedrick, 'Romancing the Salve: Sir Kenelm Digby and the Powder of Sympathy', *British Journal for the History of Science*, 41 (2008), 161–85, and Sheri L. McCord, 'Healing by Proxy: The Early Modern Weapon Salve', *English Language Notes*, 47 no. 2 (2009), 13–24.

[14] The first point may allude to some of the issues raised in Boyle's 'Salubrity of the Air' (1685), *Works*, vol. 10, pp. 303ff. The role of transplantation is discussed in *Usefulness*, *Works*, vol. 3, pp. 430–31, 433–5. See also ibid., vol. 10, p. 85.

[15] I am not aware of any discussion of this topic in Boyle's writings.

[16] See above, n. 12.

[17] Boyle's interest in such agricultural improvements is plain from certain sections of his *Usefulness of Natural Philosophy*: see esp. *Works*, vol. 6, pp. 411–14.

[18] For Boyle's active concern with this classic preoccupation of alchemists, see esp. Lawrence M. Principe, *The Aspiring Adept: Robert Boyle and His Alchemical Quest* (Princeton, NJ, 1998).

[19] A specific instance of this appears in 'Strange Reports', *Works*, vol. 11, p. 432. See also *Works*, vol. 3, p. 398, but Boyle's interest in the composition and characteristics of glass is widely in evidence throughout his corpus.

[20] In *Certain Physiological Essays*, Boyle specifically criticised scholastic philosophers for their refusal to accept that such transmutation was possible, and its investigation thus became something of a leitmotif of the mechanical philosophy: see *Works*, vol. 2, p. 91.

[21] For Boyle's interest in the alkahest of J.B. van Helmont, see esp. William R. Newman and Lawrence M. Principe, *Alchemy Tried in the Fire* (Chicago, IL, 2002), pp. 292–4.

[22] Again, this ambition relates more directly to the interests of Hooke and, in this case, Wren, than Boyle, though for his awareness of such matters, see, for example, *Works*, vol. 6, pp. 445, 517. For Wren's attempt to make hyperbolical lenses, divulged in *Philosophical Transactions* in 1669, see J.A. Bennett, *The Mathematical Science of Christopher Wren* (Cambridge, 1982), pp. 34ff.

[23] It has not been possible to throw further light on this ambition from Boyle's corpus.

The practicable & certain way of finding Longitudes.[24]

The use of Pendulums at Sea, and in Journeys, and the application of it to Watches.[25]

Potent Druggs to alter or exalt imagination, waking, Memory, and other functions and appease paine, procure innocent sleep, harmles dreams &c.[26]

A ship to saile with all winds, and a ship not to be sunk.[27]

Freedom from Necessity of much sleeping, exemplify'd by the Operations of Tea and what happens in Mad-men.[28]

Pleasing Dreams & Physicall Exercises exemplify'd by the Egyptian Electuary and by the Fungus mention'd by the French Author.[29]

Great Strength & Agility of Body, Exemplify'd by that of Frantick, Epiletick and Hystericall Persons.[30]

A perpetuall Light.[31]

---

[24]   For brief comments on the importance of such knowledge, see *Works*, vol. 11, p. 117; vol. 12, p. 505, but Boyle nowhere elaborated on this. However, for Hooke's interest in the longitude question, see esp. Michael Wright, 'Robert Hooke's Longitude Timekeeper', in Michael Hunter and Simon Schaffer (eds), *Robert Hooke: New Studies* (Woodbridge, 1989), pp. 63–118. See also Jim Bennett, 'Hooke's Instruments', in *London's Leonardo*, pp. 63–104, on pp. 68ff, and J.A. Bennett, *Mathematical Science*, ch. 5.

[25]   See the references in n. 24.

[26]   This is again a topic which has been investigated most fully in relation to Hooke. See Lisa Jardine, 'Hooke the Man: His Diary and His Health', in *London's Leonardo*, pp. 163–206, esp. pp. 181ff.; see also Lisa Jardine, *The Curious Life of Robert Hooke* (London, 2003).

[27]   For the former, Boyle's interest in collecting data from seamen in his workdiaries is perhaps relevant, though once again the topic was one that also preoccupied Hooke (see, for example, his *Posthumous Works*, ed. Richard Waller (London, 1705), pp. 561ff.). For the latter, Boyle was perhaps thinking of experiments like Sir William Petty's 'double-bottom': see the Marquis of Lansdowne (ed.), *The Double Bottom or Twin-hulled Ship of Sir William Petty* (Oxford, 1931). For Wren's interest in related matters, see Bennett, *Mathematical Science*, pp. 45–6.

[28]   See above, n. 26. Boyle had discussed the medicinal virtues of tea in *Usefulness, Works*, vol. 3, pp. 356, 449.

[29]   Neither the 'Egyptian Electuary' nor the fungus mentioned by the 'French Author' have been identified.

[30]   Another topic that I am not aware that Boyle discussed in his writings.

[31]   On the first page of his *Lampas* (London, 1677), Robert Hooke specifically disavowed any ambition to making a 'perpetual Lamp', which he dismissed as a 'Chimera', but both the gadgets he presented in that book, and that presented by Boyle in 'A New Lamp', published in *Philosophical Collections* in 1681 (*Works*, vol. 9, pp. 447–8), sought to make lamps that were as long-lasting as practicable.

Varnishes perfumable by Rubbing.[32] /fol. 78/

The Prices and other Recompences assign'd to Inventors.[33]
The Charges of hopefull Experiments borne by the Publick.[34]
A Catalogue of what is perform'd with a particular account of the wayes.
A Catalogue of *Desiderata* and Polychrista.[35]
Recompence for Appoximations [*sic*].
A Revenue for Tryalls.
A Councell for advising.[36]
A Collection of Modells, and other informative things both naturall & Artificiall.[37]

---

[32] This is precisely the kind of topic Boyle dealt with in *Usefulness*, but it does not appear therein. For a brief discussion of 'good sents produced by unlikely means', see *Works*, vol. 8, pp. 384–5.

[33] See above, pp. 26–7. Boyle is perhaps also thinking here of the contemporary patent system and the ways in which it might be improved. See Christine MacLeod, *Inventing the Industrial Revolution: The English Patent System 1660–1800* (Cambridge, 1988).

[34] This is possibly a wistful allusion to the Royal Society's mainly unrealised ambitions along such lines, as is also the case with the comparable items that occur two entries later. See Michael Hunter, *Establishing the New Science: The Experience of the Early Royal Society* (Woodbridge, 1989), passim.

[35] That is, polychrest, something adapted to several different uses (*OED*). This and the previous point reflect Boyle's concern with record-keeping in all areas of science.

[36] Again, possibly an allusion to the intended role in this respect of the Royal Society.

[37] There is an echo here of Boyle's 'Designe about Natural History': see above, n. 5.

Chapter 2

Boyle's Early Intellectual Evolution: A Reappraisal[1]

It is hardly surprising that we are all interested in 'origins stories' concerning great scientists, seeking the roots of the preoccupations that were later to make them famous, preferably as early as possible in their lives. Thus we have the collecting activities which characterised the young Charles Darwin, or the appetite for popular science books on the part of Albert Einstein.[2] Perhaps the classic example is that of Newton, whose precocious childhood inventiveness was recorded in detail by his early biographer, William Stukeley. Stukeley wrote how 'Every one that knew Sir Isaac, or have heard of him, recount the pregnancy of his parts when a boy, his strange inventions, and extraordinary inclination for mechanics.'[3] Similar stories were recorded concerning such contemporaries as John Flamsteed, Robert Hooke and Sir William Petty, the former in his autobiography, the latter in the accounts of them in John Aubrey's *Brief Lives*, where Aubrey also recorded similar proclivities on his own part.[4]

By contrast, in the case of Boyle, though the records of his childhood are unusually full, they are singularly disappointing from this point of view. It would be gratifying to be able to point to comparable nuggets of information among the profuse records that survive concerning Boyle's earliest years in the accounts and correspondence of his father, the Great Earl of Cork, or in Boyle's own autobiography, *An Account of Philaretus during his Minority*, written when he was 21 or 22, but any symptoms of an interest in the natural world in his childhood years are absent.[5] While a boy at Eton, Boyle was exposed to what has

---

[1]   Previously published as 'Robert Boyle's Early Intellectual Evolution: A Reappraisal', in special Boyle issue of *Intellectual History Review*, 25 (2015), 5–19.

[2]   See Janet Browne, *Charles Darwin: Voyaging* (London, 1995), p. 13 and passim; Roger Highfield and Paul Carter, *The Private Lives of Albert Einstein* (London, 1993), p. 16 and passim.

[3]   Quoted in F.E. Manuel, *A Portrait of Isaac Newton* (Cambridge, MA, 1968; reprinted London, 1980), p. 38. See ibid., pp. 38–41, passim.

[4]   Francis Baily (ed.), *An Account of the Revd John Flamsteed* (London, 1835–37; reprinted London, 1966), pp. 10–11; John Aubrey, *Brief Lives, Chiefly of Contemporaries*, ed. Andrew Clark (2 vols, Oxford, 1898), vol. 1, pp. 35–7, 409–10; vol. 2, p. 140.

[5]   For *Philaretus*, see *Boyle by Himself and His Friends*, pp. 1ff. In ibid., p. 9, Boyle does mention his discovery of mathematics as an antidote to the 'raving' to which he was prone

been described as 'one of the earliest English scientific libraries which is still kept together', owned by his schoolmaster, John Harrison, but the author who drew attention to this had to admit with some disappointment that 'it was only later that he began to be interested in scientific studies'.[6]

Slightly more promising is the evidence concerning Boyle's stay on the Continent between 1639 and 1644. While in Geneva, he was tutored in mathematics, especially in geometry, the doctrines of the sphere and globe and fortification, and in *Philaretus* he explained how 'enamor'd' he became 'of those delightfull studys'; the tuition that he received on such subjects is also illustrated by his recently discovered Geneva notebook.[7] More striking is the passage in *Philaretus* in which Boyle recorded how, while he was in Florence, he read not only 'the Moderne history in Italian', but also 'the New Paradoxes of the greate Star-gazer Galileo', and how he heard of Galileo's death while he was there.[8] Indeed, on the basis of this, one of Boyle's biographers, Louis Trenchard More, wrote how Galileo's writings 'deeply impressed him and very probably influenced him to adopt a scientific career', going so far as to assert that the year 1642, 'in which occurred the death of Galileo, the birth of Newton, and the beginning of Boyle's study of the new mechanics, was a memorable one in the Renaissance of science.'[9] Such views of an almost apostolic succession from Galileo to Boyle have been more or less explicitly echoed by others who want to see Boyle, like his eminent peers, as a scientist at the earliest possible date.[10]

Against this, on the other hand, is evidence that has been adduced in the past twenty years to suggest that, in the earliest phase of his intellectual career in the years following his Grand Tour, Boyle's principal preoccupation was not with science, but with writing ethical and literary works, and that it was only later – in 1649, when he was 22 – that he was to discover the excitement of

---

from his schooldays onwards, but this evidently occurred later: see ibid., pp. 14–15. On the sources associated with Cork, see *Between God and Science*, p. 264 and passim.

[6]   Robert Birley, 'Robert Boyle's Head Master at Eton', *NRRS*, 13 (1958), 104–14, on p. 107.

[7]   See *Boyle by Himself and His Friends*, pp. 14–15, and also *Between God and Science*, pp. 47, 53–6. On the discovery of the Geneva notebook, Royal Society MS 44, see L.M. Principe, 'Newly Discovered Boyle Documents in the Royal Society Archive', *NRRS*, 49 (1995), 57–70.

[8]   *Boyle by Himself and His Friends*, p. 19. Boyle also recorded looking through a telescope while in Florence: see *Works*, vol. 1, p. 88; vol. 13, p. 166. For other references to observations made in Italy of a kind consonant with the Grand Tour, see ibid., vol. 2, pp. 73, 175, vol. 3, p. 308, vol. 4, p. 361, vol. 6, p. 411, vol. 7, p. 23.

[9]   L.T. More, *The Life and Works of the Honourable Robert Boyle* (New York and London, 1944), p. 48.

[10]   For a recent example, see F.A. Buyse, 'Spinoza, Boyle, Galileo: Was Spinoza a Strict Mechanical Philosopher?', *Intellectual History Review*, 23 (2013), 45–64, on p. 56.

science that was to dominate the rest of his life.[11] It needs to be stressed here that we are not dependent on scattered hints for evidence of this, as we are in Boyle's earlier years. Instead, we have a plethora of writings by Boyle from the mid- to late 1640s and, insofar as there is evidence that further texts once existed that are now lost, these seem to have been similar in nature.[12] Yet these are not about science at all, a point made perhaps most clearly by John Harwood in his edition of the most important of them, *The Early Essays and Ethics of Robert Boyle* (1991). As he wrote in his introduction:

> I have found nothing about his childhood or adolescence that indicated a special interest in or aptitude for natural philosophy. The writings in this volume defined his early interests as the reform of moral philosophy and the pursuit of godliness.[13]

The works in question, formerly surviving only in manuscript, but now almost all in print, are profuse, and they range from ethical treatises like Boyle's *Aretology, or Ethicall Elements*, which dealt with the definition and pursuit of virtue, through religious writings such as 'Occasionall Meditations' and reflections on sin and piety, to moral essays written in epistolary form.[14] Boyle also assiduously experimented with the romance form that flourished in France at this time and was naturalised in England by his brother, Lord Broghill, in his *Parthenissa*. In Boyle's case, this is to be seen in such writings as his semi-fictional *Life of Joash*, and perhaps above all in the original version of his work, *The Martyrdom of Theodora*, which he wrote at this time. His *Account of Philaretus during his Minority* belongs to the same milieu, since it, too, is heavily influenced

---

[11]   See esp. Michael Hunter, 'How Boyle Became a Scientist', *History of Science*, 33 (1995), 59–103, reprinted in *Scrupulosity and Science*, pp. 15–57. References are here given to the version in *Scrupulosity and Science*, but for the reader's convenience, these are followed by references in parentheses to the version in *History of Science*.

[12]   See Lawrence M. Principe, 'Virtuous Romance and Romantic Virtuoso: the Shaping of Robert Boyle's Literary Style', *Journal of the History of Ideas*, 56 (1995), 377–97, on pp. 380–81, 386.

[13]   John T. Harwood (ed.), *The Early Essays and Ethics of Robert Boyle* (Carbondale, IL, 1990), p. xxiii. Cf. *Scrupulosity and Science*, p. 20 (pp. 62–3); see also the important study by Lawrence M. Principe, 'Style and Thought of the Early Boyle: Discovery of the 1648 Manuscript of *Seraphic Love*', *Isis*, 85 (1994), 247–60, which demonstrated that the references in the published version of *Seraphic Love* (1659) which had often previously been taken to indicate scientific interests on Boyle's part when he originally composed the work in 1648 had all been added at a later date.

[14]   For these writings, see Harwood, *Early Essays and Ethics*, and *Works*, vol. 13, pp. 3–144.

by the romance form.[15] In short, on the basis of the profuse output which he has actually left us, one would be forgiven for concluding that at this stage in his career, Boyle seemed set to become a moralist and litterateur, not a scientist at all.

Yet by the mid-1650s, science had come to the fore in Boyle's intellectual priorities. It was at this time that he began such seminal works as *Certain Physiological Essays* and *The Usefulness of Natural Philosophy*, both published after the Restoration, and many of his later preoccupations as a scientist can be traced back to this period in his career. What is more, it is possible to pinpoint the stage at which this change occurred, as I argued in a paper in 1995 on the basis of material in the Boyle archive that dated from the early to mid-1650s and that contrasted in subject matter with the material from the 1640s on which Harwood's verdict was based. I believe that in 1649 Boyle had a real conversion experience (and I continue to consider this Pauline metaphor appropriate, notwithstanding the sarcasm with which some have received it).[16] The most striking evidence for this is provided by Boyle's workdiaries, which are now available online, the paperbooks in which he made entries from 1647 until the end of his life.[17] In format, these show a substantial continuity, often with titles like 'Diurnall Observations, Thoughts & Collections. Begun at Stalbridge April 25th 1647' or 'A Private Philosophicall Diary Begun this first day of January 1655/6'. Yet at a specific date, in 1649, the subject matter changes dramatically and irreversibly from moral aphorisms and literary extracts to recipes and observations concerning chemical processes. The break occurs between a workdiary dated 25 March 1649 and the next, dated New Year's Day 1650: up till then we have lengthy extracts from French romances and the like, but from then onwards the workdiaries are made up exclusively of recipes and laboratory data, and ethical and literary aphorisms never recur. The most plausible explanation for this is that there was a sudden and dramatic change in Boyle's interests. Insofar as there is any alternative, perhaps of claiming that there must have been some earlier scientific workdiaries that are now lost (or, reciprocally, some later literary ones, also oddly missing), there is no evidence of this, and it smacks of special pleading.

Then there are the works Boyle began to write, starting with one entitled 'Of the Study of the Booke of Nature'.[18] Again, they are all now in print, and they are extraordinarily different from the crabbed ethical treatises of the 1640s or even

---

[15]   For all this, see *Boyle by Himself and His Friends*, esp. pp. xvff., and *Between God and Science*, ch. 4.

[16]   See Charles Webster, 'Introduction to the Second Edition', in Webster, *The Great Instauration* (2nd edn, New York and Bern, 2002), p. xxxix, referring to *Scrupulosity and Science*, p. 3. See also above, n. 11.

[17]   See http://www.livesandletters.ac.uk/wd/index.html.

[18]   *Works*, vol. 13, pp. 145–72.

the variations on the romance form with which Boyle experimented towards the end of the decade. 'Of the Study of the Booke of Nature' shows a new, gushing enthusiasm for the evidence of God's design as revealed in the universe, and the potential for promoting piety by studying this. What is more, the writing of this work seems to dovetail with references in Boyle's correspondence which bear out the fact that there was a change in his intellectual priorities at this point, even insofar as he continued to complete commitments that he had already undertaken. Thus he had to apologise in the introduction to 'An Invitation to Communicativenesse' – dated 23 July 1649 and evidently addressed to Samuel Hartlib – that the time for work on it had to be 'snatcht from my newly-erected Furnaces', while more revealing still is Boyle's letter of 31 August that year to his sister, Lady Ranelagh (who may well have been a party to the transformation), in which he explained to her that 'Vulcan has so transported and bewitch'd mee, that ... [I] fancy my laboratory a kind of Elizium.' Equally notable is his reference in the same letter to 'those Morall speculations, with which my Chymicall Practises have entertained mee', and to a discourse on 'the Theologicall Use of Naturall Filosophy' – evidently 'Of the Study of the Booke of Nature' – with which he would by then have presented her if illness had not impeded this.[19] Clearly something new and exciting had happened, and, in conjunction with the archival evidence I have outlined, it should be clear to any but the most obtuse what it was.

What is more, there is a continuity in Boyle's preoccupations from this point onwards, in terms of his commitment to the study of nature by experimental means and his conviction of both the intellectual and religious importance of the task. A clear progression is in evidence through identifiable phases of development, including his encounter with George Starkey and then with Oxford thinkers like Ralph Bathurst, followed by Boyle's profuse activity in the late 1650s in writing the treatises on natural philosophy which were to be published to widespread acclaim after 1660. The rest of his intellectual career was in many ways a sequel to this, though with further developments, to all of which I have tried to do justice in my biography, *Boyle: Between God and Science*.[20]

It is also worth drawing attention here to Boyle's parallel discovery of erudition in the years around 1650, the significance of which should not be underestimated just because we 'know' that Boyle's destiny was to become known primarily as an influential experimental philosopher. The significance of this is clear partly from comments that Boyle later vouchsafed to Bishop Gilbert Burnet about the influence on him at this point of the great scholar Archbishop

---

[19] Ibid., vol. 1, p. cxiii; *Correspondence*, vol. 1, pp. 82–3.

[20] See *Between God and Science*, chs 5–7 and passim. For the key role of Starkey, see also William R. Newman and Lawrence M. Principe, *Alchemy Tried in the Fire: Starkey, Boyle and the Fate of Helmontian Chymistry* (Chicago, IL, 2002).

James Ussher, and partly from another key treatise of this transitional period, Boyle's 'Essay of the Holy Scriptures', in which he made a powerful case for knowledge of the biblical languages and of the background to the Bible in defending the Christian dispensation against attack.[21] This has now been freshly illustrated by the attention David Cram has given to a document in the Boyle Papers which I had long known about, but had never felt equipped to interpret.[22] This proves to be a fragment of the Hebrew grammar that Boyle told Burnet he compiled for his own use at this time, commenting on aspects of Hebrew grammar from a comparative philological point of view: it notes differences between Hebrew, Syriac, Chaldean and other forms, and specifically discusses the role of affixes in verbal conjunctions as the key to the grammatical system. It is a dense and sophisticated piece of work, based on perusal of recent scholarly books by Ludovicus de Dieu, Johann Ernst Gerhard and Victorinus Bythner.[23] Hence this 'new' document bears out the seriousness of such commitments at that stage in Boyle's career more clearly than ever, again raising the question of the inevitability of his trajectory towards science, since he might have become a learned biblical exegete instead.

Why had the change occurred, and how did it relate to developments in what was, after all, a traumatic period of English history, with the Civil War and its aftermath? Though certain commentators have attributed great significance to the impact of the war on Boyle's intellectual development, as we will see shortly, I am sceptical about crude formulations of such claims since, at the time when he wrote his ethical treatises, Boyle had been almost oddly detached from the political developments of the period. The context of his conversion experience was arguably equally intellectualist. He now became aware of a threat, not from the post-war breakdown of religious control, but from an attack on religion altogether, as represented by sixteenth-century thinkers like Machiavelli and Pomponazzi. This was a threat which he saw as abetted by the prevailing Aristotelianism and which he thought that the new philosophies, especially that of the 'chymists', could help to overcome. Of course, he attached a similar importance to erudition, a further tool by which the enemies of Christianity might be disarmed. As to why Boyle should have had this awakening at this

---

[21]   See *Boyle by Himself and His Friends*, p. 27; *Works*, vol. 12, pp. 55–6, vol. 13, pp. 173–223. See also *Scrupulosity and Science*, pp. 32–5 (pp. 71–3), and *Between God and Science*, pp. 79ff.

[22]   BP 41, fols 64–5. Adjacent to this is a set of mnemonic verses relating to Latin phonetics and metres (fols 66–7), the date and significance of which are unclear. I am grateful to David Cram for sharing his provisional findings on this document with me.

[23]   The works in question are de Dieu, *Grammatica linguarum orientalium* (Leiden, 1628); Gerhard, *Institutiones linguæ Ebreæ* (Jena, 1647) and Bythner, *Lingua eruditorum* (Oxford, 1638). Interestingly, the latter two were recommended by Boyle to John Mallet in January 1653: *Correspondence*, vol. 1, pp. 140–41.

moment, there are various possibilities which I canvassed in my 1995 article, including his exposure to the intellectual culture of Amsterdam on a visit in 1648, and possibly even the national trauma over the execution of Charles I in January 1649 – though I wonder whether a key factor was the influence of Ussher, who is known to have been concerned about sceptical trends, not least as represented by the proto-Deist Lord Herbert of Cherbury.[24] I am not sure that any such stimulus can be claimed to decisive in itself, but this does not reduce the significance of the change that occurred.

I will return in due course to this claim for a momentous turning point in 1649, but first I want to consider various accounts that have been given of Boyle's intellectual evolution, starting with certain documents giving 'origins stories' on the part of Boyle himself, and then examining the views of later commentators. In fact, there are two little-known texts by Boyle which give rather interesting such narratives, in both cases seemingly indicating an earlier proclivity for the study of natural philosophy than one might otherwise expect. One is the 'Introductory Preface' to an otherwise lost work written by Boyle in the mid-1650s which opens with the statement that, 'Having ever since I first addicted my Time & Thoughts to the more serious parts of Learning, found my selfe by a secret but strong Propensity, inclin'd to the Study of Naturall Philosophy', his first study was of the accepted Aristotelian doctrine:

> But I had scarce well acquainted my selfe with the Peripatetick Theory, before I was strongly tempted to doubt it's Solidity. For the commands of my Parents engaging me to visit divers forreigne Countrys, I could not but in my Travells, meet with many things capable to make me distrust the Doctrine wherewith I had freshly been imbu'd.

He went on to explain how he became devoted instead to philosophers like Telesio, Campanella, Gassendi and Descartes.[25] However, it is hard to put much faith in the document in question, which is rather schematic and clichéd. It is wholly implausible that Boyle had become as adept an Aristotelian as is there claimed by the time he went to the Continent at the age of 12½, while the invocation of his 'parents' as the source of the instruction that he should travel abroad is also suspicious, since all decisions concerning the young Boyle were made by his father after his mother died when Boyle was aged three. It reads

---

[24] *Scrupulosity and Science*, pp. 42ff. (pp. 79ff.). The possible influence of Hartlibians like Worsley (on whom, see below), is also there discussed.

[25] BP 38, fol. 80, printed in M.B. Hall (ed.), *Robert Boyle on Natural Philosophy* (Bloomington, IN, 1965), pp. 177–9. The MS is damaged, and the name that Hall reads as 'Ba[con]' is in fact probably 'Bas[so]'.

like a just-so story, and it is doubtful whether it can be used to prove very much about Boyle's early intellectual evolution in the absence of corroborative data.[26]

Equally interesting is another text, a section of Boyle's 'Essay of the Holy Scriptures' which is only known from the version of it printed by Thomas Birch in the 'Life of Boyle' prefixed to the edition of Boyle's *Works* that Birch brought out in 1744. This means that there is no handwriting evidence concerning its date, but it seems likely that this, too, dates from the mid-1650s. Looked at in the current context, this also seems rather schematic, though it is not without interest. In this text, Boyle is setting up a rhetorical contrast between 'words' and 'things': he thus speaks of 'those excellent sciences, the mathematics, having been the first I addicted myself to, and was fond of, and experimental philosophy with its key, chemistry, succeeding them in my esteem and applications', by way of illustrating how 'my propensity and value for real learning gave me so much aversion and contempt for the empty study of words' that he neglected linguistic studies. He went on:

> But in spite of the greatness of these indispositions to the study of tongues, my veneration for the scripture made one of the greatest despisers of verbal learning leave *Aristotle* and *Paracelsus* to turn grammarian, and ... learn as much Greek and Hebrew, as sufficed to read the old and new testament ... and thereby free himself from the necessity of relying on a translation.[27]

He then proceeded to describe his more advanced studies in biblical languages, in Chaldean and Syriac. Commentators have sometimes cited this passage as telling evidence of Boyle's conversion from the mathematics to which we know he was exposed in his adolescence to 'experimental philosophy with its key, chemistry' – all of which was supposed to have occurred before his encounter with erudition at the hands of Archbishop Ussher and others in the late 1640s.[28] But surely what is significant here is the rhetorical point he was seeking to make about his late conversion to biblical learning, which makes me very suspicious of his comments about his ostensibly earlier studies.

From Boyle's later years, we have disappointingly few accounts of his intellectual evolution, the main exception being the 'Burnet Memorandum', the notes Boyle dictated to Bishop Gilbert Burnet around 1680 in connection with Burnet's plan to write a life of Boyle, though this never materialised. This does indeed contain a revealing rationalisation of Boyle's later scientific programme, in which he prioritised *The Usefulness of Natural Philosophy* and explained how

---

[26]   See the discussion in *Scrupulosity and Science*, pp. 21n., 26n. (nn. 25, 47–8).

[27]   *Works*, vol. 12, pp. 356–7.

[28]   See, for example, J.J. O'Brien, 'Samuel Hartlib's Influence on Robert Boyle's Scientific Development', *Annals of Science*, 21 (1965), 1–14, 257–76 (published in 1966), on p. 2.

various subsequent works developed from that.[29] On the earlier development of Boyle's scientific interests, on the other hand, it is disappointingly sparse (in contrast to its striking comments about his commitment to biblical study under Ussher's aegis that have already been cited): the main comment comes in connection with his stay in Geneva, where Boyle states: 'he fell on Senecas Naturall questions which first set him on Naturall Philosophy'.[30] It is a little difficult to know what to make of this in the absence of evidence of study of natural philosophy on Boyle's part till much later, though it is perhaps to be read in the context of an adjacent comment indicating Boyle's exposure to Stoic doctrine at this point; it is perhaps also revealing that Seneca's *Quaestiones naturales* was the most moralistic of all ancient writings on such subjects, and that it was in this connection that Boyle was later to cite the work.[31]

This means, I have to admit, that Boyle himself never gave a very explicit account of the turning point I outlined earlier as far as experiment, as against erudition, is concerned. However, there is one contemporary testimony that could be interpreted as doing so: the dedication to Boyle of the *History of Generation* (1651) by Nathaniel Highmore, Boyle's Dorset neighbour in the early 1650s and himself a notable medical researcher – in fact, the first book to be dedicated to Boyle. For this comprises two paragraphs, and, whereas in the first Highmore comments how Boyle's pursuit of virtue acted as 'a pattern and wonder' to his benighted peers, in the second he goes on to praise Boyle for his studies of 'Nature in her most intricate paths', and for not considering his blood and descent debased by these. Indeed, though in his statement that Boyle had 'made a better and far nobler choice' he might be seen as comparing him with his peers, he might also have intended to make a contrast between Boyle's former ethical concerns and his new-found interest in the study of nature.[32]

Turning to later biographical accounts of Boyle, these are, of course, dominated by the presumption that he was destined to become a scientist at the earliest possible stage, but it is interesting to see what such commentators saw as the formative phases in this process. It is appropriate to start with Thomas Birch's 1744 'Life of Boyle', probably the most influential biographical account of the great man ever written. On Boyle's early years, Birch simply printed *Philaretus* for the first time, while to illustrate his development in the late 1640s, he quoted the text of various letters verbatim. On the basis of these, he drew attention to two key developments at this point in Boyle's career which have dominated interpretation ever since. One was Boyle's link with the intelligencer Samuel

---

[29]   *Boyle by Himself and His Friends*, pp. 28–9.
[30]   Ibid., p. 26.
[31]   See the discussion in *Between God and Science*, pp. 49–50.
[32]   Highmore, *History of Generation* (London, 1651), sigs ¶3–4. I am grateful to Peter Anstey for discussion of this text.

Hartlib, who was seen as stimulating Boyle's interest in natural philosophy. The other was his association with the so-called Invisible College, which is mentioned in three of the letters Birch printed, and which he identified as the group of natural philosophers who met in London in 1645, as described by the mathematician John Wallis in a famous account that Birch also published at length. This group, in Birch's words, 'applied themselves to experimental inquires, and the study of nature, which was then called the new philosophy, and at length gave birth to the Royal Society', so it seemed only appropriate that Boyle's own scientific apprenticeship should occur in this milieu.[33]

In fact, of course, Birch was wrong in identifying the 1645 London group with the Invisible College, and, though he was followed in doing so by such biographers as Fiona Masson and L.T. More, thereafter matters were set straight both by historians investigating the origins of the Royal Society like Rosemary Syfret and Douglas McKie, and by such Boyle scholars as Marie Boas and R.E.W. Maddison, all of whom pointed out that the Invisible College could not be the 1645 group, but must be something else, which they tended to associate more or less closely with the Hartlib circle.[34] We will return to the Invisible College in due course, but here it is worth noting that Birch's identification has arguably left a subconscious legacy among subsequent commentators, who, though abandoning his view that the two were the same, have nevertheless often retained a presumption that the Invisible College must have partaken of a similar agenda to that of the 1645 group.

The stress on the significance for Boyle of the Hartlib circle to be found in writings like these was then taken up by J.J. O'Brien in a pair of articles published

---

[33]   Thomas Birch, 'The Life of the Honourable Robert Boyle', in Birch (ed.), *The Works of Robert Boyle*, 2nd edn, 6 vols (London, 1772), vol. 1, pp. vff.; the quotation is from ibid., p. xlii. In addition to *Philaretus*, Birch briefly quoted the Burnet Memorandum on Boyle's interest in Stoicism (though not the specific passage quoted above): ibid., p. xxvi. For Birch's comments on Hartlib, see ibid., p. xxxvii; for the Invisible College references, see ibid., pp. xxxiv–xxxv, xl. See also Birch, *Royal Society*, vol. 1, p. 2.

[34]   See Flora Masson, *Robert Boyle: A Biography* (London, 1914), pp. 148–9; More, *Life and Works of Boyle*, pp. 54, 64–5; a later reassertion of the identification appears in G.H. Turnbull, 'Samuel Hartlib's Influence on the Early History of the Royal Society', *NRRS*, 10 (1953), 101–30, on pp. 102–3, 129. For the contrasting view, see Rosemary Syfret, 'The Origins of the Royal Society', *NRRS*, 5 (1948), 75–137, on pp. 119–29; Marie Boas, *Robert Boyle and Seventeenth-century Chemistry* (Cambridge, 1958), pp. 5–7, 31–3 and ch. 1, passim; Douglas McKie, 'The Origins and Foundation of the Royal Society of London', *NRRS*, 15 (1960), 1–37, on pp. 21–3; R.E.W. Maddison, 'Studies in the Life of Robert Boyle, F.R.S. Part VI: The Stalbridge Period, 1645–55, and the Invisible College', *NRRS*, 18 (1963), 104–24, on pp. 109–11; R.E.W. Maddison, *The Life of the Hon. Robert Boyle* (London, 1969), pp. 67–9; Margery Purver, *The Royal Society: Concept and Creation* (London, 1967), pp. 193–205.

in *Annals of Science* in 1966 on 'Samuel Hartlib's Influence on Robert Boyle's Scientific Development', which give a full if slightly imprecise account of Boyle's relations with Hartlib, the implication being that it was Boyle's association with Hartlib which initially introduced him to the interests they shared.[35] The Hartlib circle also figures prominently in the interpretation of Boyle by J.R. Jacob, initially divulged in various articles and then published as *Robert Boyle and the English Revolution* in 1977, in which the influence of Hartlib and his friends is presented as a major catalyst in Boyle's intellectual evolution.[36] This forms part of a broader claim on Jacob's part – perhaps the boldest thesis concerning Boyle's development at this time – namely that the Civil War and its aftermath had a profound effect on Boyle, making him question the traditional social ethic with which he had been brought up, and instead advocate a programme of the pursuit of individual virtue and public profit. The most influential part of Jacob's argument concerned the supposed evolution of Boyle's view of nature: he claimed that Boyle engaged in a 'dialogue' with the radical sects which emerged after the Civil War, and that not only his social ethic, but also his corpuscular philosophy was consciously forged as a reaction to their vitalist view of nature. In other words, Boyle's view of the universe as comprising inert particles of matter presupposed that their activity derived from an external God, thus providing support for existing political and religious hierarchies, whereas the sectaries' claims that matter was itself active had subversive levelling implications.

This formed part of a broader view of the science in the period as essentially counter-revolutionary which Jacob developed in conjunction with his then wife, Margaret Jacob, and which has proved extraordinarily influential, even though it has to some extent been superseded by the more sophisticated claims along similar lines in Steven Shapin's and Simon Schaffer's *Leviathan and the Air-pump* (1985).[37] Arguably the thesis always appealed particularly to political historians who were delighted to find scientific developments mirroring political ones as closely as the Jacobs affirmed and who were ill-equipped to assess the validity of such claims for themselves. In fact, as I have illustrated elsewhere, J.R. Jacob's view of a 'dialogue with the sects' seriously misrepresents the evidence, and his view of Boyle's science as counter-revolutionary is also highly questionable: quite apart from the fact that we now know that matter theory was not polarised

---

[35]   O'Brien, 'Samuel Hartlib's Influence'.

[36]   J.R. Jacob, *Robert Boyle and the English Revolution* (New York, 1977), pp. 16ff. and passim. Of the earlier articles, see especially J.R. Jacob, 'The Ideological Origins of Robert Boyle's Natural Philosophy', *Journal of European Studies*, 2 (1972), 1–21.

[37]   See esp. their joint article, 'The Anglican Origins of Modern Science: the Metaphysical Foundations of the Whig Constitution', *Isis*, 71 (1980), 251–67. For a contextualised view of the impact of Jacob and of Shapin and Schaffer, see Michael Hunter, 'Scientific Change: Its Setting and Stimuli', in Barry Coward (ed.), *A Companion to Stuart Britain* (Oxford, 2003), pp. 214–29.

between conservative mechanists and radical vitalists in the manner he asserted, it also turns out that Boyle was surprisingly sympathetic to the views of the very religious radicals to whom, according to the Jacob thesis, he should have been opposed.[38] I should perhaps add that Jacob was not aware of or interested in the transition in Boyle's interests that I have since claimed occurred in 1649, since natural philosophy and ethics form a seamless whole in his view of things (and erudition hardly figures at all).

I now turn to a view of Boyle's early intellectual career that has proved more influential, due not least to the magisterial scale of the book in which it was put forward, Charles Webster's *The Great Instauration: Science, Medicine and Reform, 1626–1660*, first published in 1975, of which a second edition appeared in 2002. In this Boyle appears as an integral member of the Hartlib circle, and, particularly in the sections dealing with 'The Prolongation of Life' and 'Dominion over Nature', Boyle and Hartlib are often treated as closely related figures. In terms of how Boyle's scientific interests originated, on the other hand, Webster is in fact subtler than either O'Brien or Jacob, laying more stress on the role of the Invisible College (to which I will return shortly), which he sees as crucial in this respect, and Boyle's association with which he sees as slightly pre-dating his contact with Hartlib himself. Indeed, it has to be admitted that the exact circumstances of Boyle's initial encounter with Hartlib are obscure, largely because of a lacuna in Hartlib's otherwise highly informative 'Ephemerides' from 1643 to 1648.

It is worth clarifying here (since the 'Introduction to the Second Edition' of *The Great Instauration* suggests that Webster is in some doubt about this[39]) that I have never for a moment doubted the significance of Boyle's close liaison with Hartlib, evidenced particularly through the lengthy correspondence between the two men from 1647 to 1659 – even if it is unfortunate that very few of Boyle's letters to Hartlib survive as against the lengthy missives that Hartlib penned to him.[40] From these, the extent of their shared interests is clear, and

---

[38]   See John Henry, 'Occult Qualities and the Experimental Philosophy: Active Principles in Pre-Newtonian Matter Theory', *History of Science*, 24 (1986), 335–81; *Scrupulosity and Science*, pp. 8, 51ff. (pp. 86ff.), 63–4, and Malcolm Oster, 'Virtue, Providence and Political Neutralism: Boyle and Interregnum Politics', in Michael Hunter (ed.), *Robert Boyle Reconsidered* (Cambridge, 1994), pp. 19–36. However, I should note here my reservations about Oster's alternative reading of Boyle's development in the 1640s as set out in his 'Biography, Culture and Science: the Formative Years of Robert Boyle', *History of Science*, 31 (1993), 177–226: see *Scrupulosity and Science*, pp. 23–4 (p. 65), and *Between God and Science*, p. 317 n. 74.

[39]   Webster, *Great Instauration*, p. xxxix. It is interesting to contrast the rather fuller and more balanced exposition of my views in his 'La Reinvenzione di Robert Boyle', *Rivista Storica Italiana*, 109 (1997), 298–306.

[40]   See *Correspondence*, vol. 1, passim.

the same is true of the references to Boyle in Hartlib's 'Ephemerides', as I hope I have fully made clear in *Boyle: Between God and Science*.[41] I am certainly not to be aligned with Rupert Hall and his 1972 essay 'Science, Technology and Utopia in the Seventeenth Century', in which he took a rather negative view of the role of the Hartlib circle in relation to the development of science in mid-seventeenth-century England.[42] On the other hand, I *do* think that there is a danger of misunderstanding Boyle by conflating him too much with the Hartlib circle and failing to do justice to the slightly idiosyncratic way in which his interests developed. Though, of course, this raises the question of just how homogeneous the Hartlib circle really was, Boyle's intellectual priorities were certainly significantly different from those of key players in it like Benjamin Worsley and Sir Cheney Culpeper.[43]

I should add that the same is true of the Oxford group with which Boyle was associated from 1655 onwards, as convened by John Wilkins of Wadham College, which evidently represents the intellectualist alternative with which Charles Webster seems to think that I want to align Boyle. In fact, I consider that Boyle's intellectual priorities can be misrepresented just as much by conflating them too closely with those of the Oxford group – as they have been in a recent study of contrasting traditions in the natural theology of the period by Scott Mandelbrote.[44] In this, Boyle is presented almost as a clone of Wilkins, and again, the result is to efface Boyle's individuality and hence to fail to do justice to the eclecticism which helps to explain why he was so influential. Mandelbrote thus fails to take account of the extent to which (like many of the Hartlibians) Boyle was an outsider to the academic culture of which Wilkins was a product; he also fails to do justice to the extent to which Boyle was sympathetic to the rival apologetic tradition associated with thinkers of a more mystical hue, including radical figures to whom, as Mandelbrote shows, Wilkins and certain of his colleagues were implacably opposed.[45] It is not that I am questioning the overall contrast between two apologetic traditions that Mandelbrote postulates,

---

[41] *Between God and Science*, chs 4–7.

[42] A.R. Hall, 'Science, Technology and Utopia in the Seventeenth Century', in Peter Matthias (ed.), *Science and Society 1600–1900* (Cambridge, 1972), pp. 33–53.

[43] On Worsley, see *Scrupulosity and Science*, pp. 46–7 (pp. 82–3); much the same may also be said for Culpeper, on whom, see Webster, *Great Instauration*, pp. 67ff., 369ff., and Stephen Clucas, 'The Correspondence of a 17th-century "Chymicall Gentleman": Sir Cheney Culpeper and the Chemical Interests of the Hartlib Circle', *Ambix*, 40 (1993), 149–70. Webster's view of the internal working of the Hartlib circle is also undercut by the analysis of the circle's impact on the young Boyle given in Newman and Principe, *Alchemy Tried in the Fire*, esp. pp. 207–72.

[44] Scott Mandelbrote, 'The Uses of Natural Theology in Seventeenth-century England', *Science in Context*, 20 (2007), 451–80.

[45] See *Scrupulosity and Science*, esp. pp. 56–7 (p. 91) and 223ff.

which seems to me substantially valid: the point is that Boyle presents a problem for it since he occupies so much of an intermediate position between them. Yet, as we shall see, that is partly the key to his significance.

I now wish to turn to the notorious Invisible College with which Boyle was associated, as documented by a handful of brief references in the mid-1640s. Not only has Charles Webster argued that this episode 'provided the occasion for Boyle's first serious excursion into science', he has also populated this elusive grouping with an elaborate Anglo-Irish caste of actors.[46] In fact, various things should be made clear. The first is that all we know about the membership of this group is that it cannot have included those to whom Boyle reported on it in letters, namely Hartlib, Isaac Marcombes (Boyle's former tutor) and a Cambridge don, Francis Tallents. Secondly, all that is known about its remit comes from Boyle's brief comments. In the one that is quoted most often, its agenda is described as comprising 'natural philosophy, the mechanics, and husbandry' on the grounds that knowledge was only valued 'as it hath a tendency to use'. Boyle also commented on its members' respect even for 'the meanest, so he can but plead reason for his opinion', on the fact that 'that school-philosophy is but the lowest region of their knowledge', and on their concern for 'the whole body of mankind'.[47] On the other hand, when discussing what is evidently the same body in his 'Doctrine of Thinking', a treatise on meditation dating from this period, Boyle attributed 'the better part of their rare and New-coynd Notions' to 'the Diligence and Intelligence of their Thoughts', and it is worth noting that the only time he uses the word 'experiments' in relation to the group is to describe his experience of these 'New-coynd Notions': 'the Experiments of this I have lately Seen in those I have had the Happiness to be acquainted with of the Filosophicall Colledge'.[48] (It is also worth noting that Boyle states that the natural philosophical and applied pursuits in which he indulged with the college were 'the *other* humane studies' to which he devoted himself 'in my spare hours' when he was not writing the ethical and literary works described earlier.[49]) However, this episode is clearly going to run and run, so let me comment on it further.

First, as I stressed when writing about Boyle's early intellectual development in my article of 1995, we need to beware of the presumption that Boyle's trajectory was preordained at this point – in other words, that what we are

---

[46]   Webster, *Great Instauration*, p. 57 and pp. 57–67, passim, and Charles Webster, 'New Light on the Invisible College: The Social Relations of English Science in the Mid-seventeenth Century', *Transactions of the Royal Historical Society*, 5th series 24 (1974), 19–42.

[47]   For the passages in question, see *Correspondence*, vol. 1, pp. 42, 46 and 58.

[48]   Harwood, *Early Essays and Ethics*, p. 186.

[49]   *Scrupulosity and Science*, pp. 22–3 (pp. 64–5) (the italics are mine). See also *Between God and Science*, pp. 66–7.

looking for is the origins of his concern with science in a modern sense: there is something deeply worrying about Webster's repeated invocation of 'experimental philosophy' in connection with the college when we actually know so little about its remit.[50] Equally important, we need to take more account than hitherto of the implication of the patchiness of our biographical knowledge about Boyle in this period in his life, and in particular of the extent to which letters to and from him in these years are known once to have existed which are now lost: this must surely affect speculation based purely on what we know about the recipients or senders of letters that happen to survive. Here, let me refer to the research that went into the 2001 edition of Boyle's *Correspondence*, which showed that in 1646 and 1647, the two years from which date the references to the Invisible College, 15 letters happen to survive, but these are greatly outnumbered by 35 that are lost.[51] Unfortunately, all we know about most of the lost letters is that they were from Boyle to correspondents who are not identified; but occasionally we are given names, as of William Strode, a Somerset parliamentarian whom no one has ever canvassed as a member of the Invisible College, while it is also worth noting here a further network of contacts of Boyle's in the late 1640s, this time in Cambridge, including Samuel Collins, the ejected Provost of King's College, whose panegyric of Boyle has recently been discovered.[52]

In the light of this, and in the absence of any real evidence as to who actually comprised the college, it seems to me that only the most arrant of special pleading makes it likely that the group comprised those with whom Charles Webster was familiar through his research on Hartlib's associates. In particular, it should be made clear that the actual references to the college and its concerns provide no mandate at all for Webster's presumption of an Anglo-Irish connection.[53] Moreover, though of its putative members Webster seems to attach special significance to Benjamin Worsley, even going so far as to describe the body as 'Worsley's Invisible College', largely on the basis of the supposed similarity of content of two letters from Boyle to Worsley that *are* extant to the vague agenda associated with the college, I personally am far from convinced that the similarity is great enough to form the basis of such claims.[54] Instead, the

---

50   Webster, 'New Light', pp. 33, 34, 37, 42.

51   See *Correspondence*, vol. 1, pp. 29–64. See also ibid., pp. xxviii–xxx.

52   Ibid., vol. 1, pp. 44, 52, 56; see Michael Hunter and David Money, 'Robert Boyle's First Encomium: Two Latin Poems by Samuel Collins (1647)', *The Seventeenth Century*, 20 (2005), 223–41.

53   It is perhaps worth noting that, though generally agreeing with Maddison over the personnel of the college, Webster rejected Maddison's inclusion of Dury as a member: *Great Instauration*, p. 59n.

54   Ibid., p. xxxviii; 'New Light', pp. 26–7. See *Correspondence*, vol. 1, pp. 42–4, 47–9. For later letters from Worsley to Boyle, see ibid., 241–3 (and pp. 208, 247 for two lost letters from Boyle to Worsley), and below. For background, see Thomas Leng, *Benjamin Worsley*

college could easily have had a completely different personnel not represented among Boyle's sparse extant correspondence from this period at all.

Now I should make it clear that I have no particular axe to grind in issuing these caveats. In undertaking the re-examination of this phase in Boyle's career from which they stem, I had no preconceived agenda: I was simply curious to find out more about Boyle and the way in which his interests and proclivities developed during the period in question. Indeed, what perhaps surprised me most was the discovery of how much time and effort he devoted in the early 1650s to linguistic erudition under the influence of Archbishop Ussher. It is true that my preoccupation with Boyle himself may have shielded me from some of the background noise that made Webster and others want to conflate his interests more closely than was warranted with those of the Hartlib circle. What was crucial, however, was my newly meticulous study of the profuse extant material dating from this period in Boyle's career, deploying a chronological precision that made it possible to examine the documents virtually on a year-by-year basis rather than lumping the whole of Boyle's 'Stalbridge' period together, as had often been done in the past.[55] It was this that led to the identification of the striking change in 1649 that I outlined in my 1995 article on 'How Boyle Became a Scientist'.

On the other hand, I was careful not to overstate the case, since, despite the decisive change that occurred in Boyle's overall preoccupations, there were undoubted elements of continuity. As we have already seen, Boyle does seem to have gained some acquaintance with the new science and become aware of the shortcomings of scholastic natural philosophy during his studies on the Continent in the early 1640s, particularly in relation to mathematics and cosmology. This evidently gave him a degree of familiarity with recent developments which occasionally surfaces in the late 1640s, for instance when he discussed authors like Gassendi and Mersenne, or the rival claims of the Ptolemaic, Tychonian and Copernican cosmologies in his earliest letters to Hartlib in 1647. There was also a fruitful intersection between the moralistic concerns that preoccupied him at this time and Hartlib's promotion and dissemination of public-spirited enterprises through his 'Office of Address'.[56] In addition, Boyle was clearly interested in the technological and agrarian improvements that he discussed in letters with Hartlib and Worsley, and in chemical topics about which he may have learned from the latter, and in 1647 he even seems to have attempted to set up a laboratory at his country house in Stalbridge, though the extant evidence

---

(1618–77): Trade, Interest and the Spirit in Revolutionary England (Woodbridge, 2008), esp. ch. 5.

[55]   See, for example, Maddison, 'Studies in the Life of Robert Boyle'.

[56]   *Correspondence*, vol. 1, esp. pp. 53, 55, 59–60, and see the commentary in *Scrupulosity and Science*, pp. 21–2 (pp. 63–4), and *Between God and Science*, p. 66.

suggests that little came of this.[57] Perhaps revealingly, very occasional allusions to matters about which he might have learned from Worsley and Hartlib *do* appear in Boyle's ethical writings in 1647 – in other words, just at the point when this might have been expected from what we know of his association with them.[58] Also relevant here is Boyle's first published writing, his 'Invitation to a Free and Generous Communication of Secrets and Receits in Physick', which Hartlib was to publish in 1655, which deals with the morality of secrecy, though arguably more from the viewpoint of a consumer of medications than a producer of them.[59] This evidence is significant, but the point is that what we end up with (in the words of my 1995 article) is a Boyle who 'was at best a reader of scientific books with a generalised curiosity about low level technology', but who 'had not yet discovered the overriding enthusiasm for experiment which led to the classic works' that were to be published after 1660. There is no question that at this point, his chief preoccupation was writing moral and religious treatises, and it was for this that he was known to Hartlib and his associates, like John Hall, who wrote to Hartlib asking him to obtain from Boyle 'A draught of those opinions of his about vertue & the ways of teaching it'.[60]

Equally, on the other hand, I would be the last to deny that there was some legacy from Boyle's preoccupations in the 1640s to the 1650s and beyond. In particular, Boyle's overriding religious and moral imperative is there both before and after the change in 1649, simply being transposed into a new area, in that it was now the natural world to which he appealed for the moral lessons that he wanted to teach his audience.[61] There is an even more direct progression in his devotional writings, including the *Occasional Reflections* that he had begun to compose in the 1640s, to which he continued to add thereafter prior to producing a published version of them in 1665: in fact, 'Of the Study of the Booke of Nature' is described in its subtitle as being linked to this work.[62] But I do not think there can be any question that the element of change was more significant than the element of continuity. Indeed, in this connection it is also worth remembering Boyle's commitment to erudition, to which I have already

---

[57] See esp. *Correspondence*, vol. 1, pp. 49–50, and the commentary in *Between God and Science*, pp. 70–71.

[58] See the allusion to 'Mercury (that Proteus that is not more Various in it's Shapes then Constant to it's Nature)' in a moral epistle dated 15 April 1647 (*Works*, vol. 13, p. 57). See also the reference to Sennert's *Institutions of Physick* in another dated 15 August 1647 (ibid., p. 70), and the chemical metaphors used in Boyle's letter to Worsley of late February 1647 (*Correspondence*, vol. 1, p. 48).

[59] See below, p. 132.

[60] *Scrupulosity and Science*, pp. 22–3 (pp. 64–5).

[61] For important comments on this theme, see Principe, 'Virtuous Romance', pp. 392ff.

[62] See *Works*, vol. 5, pp. xi–xv and 3–187, and vol. 13, pp. xxx–xxxii, xxxvii, 101–16 and 147.

alluded, which was clearly a further part of the shift in his priorities that occurred around 1650. This was significant not only in itself, but also, as I have argued elsewhere, for the extent to which it contributed to his mature style.[63] This too, therefore, only adds to the significance of the transition which seems to have occurred at a specific time, an appreciation of which I consider crucial to our understanding of Boyle.

Of course, some have remained unconvinced, including Antonio Clericuzio, who has questioned my claim in various recent writings. It therefore seems worthwhile to run through his criticism here and to illustrate why, in my view, it is unfounded. 'To start with,' he says, 'the fact that before 1649 Boyle wrote on ethics does not mean that he was not pursuing scientific research. There is no reason to assume that writing tracts on ethics and pursuing natural philosophy were mutually exclusive occupations.'[64] But I have never been guilty of any such presumption: my claim was based on contrasting the subject matter of 'Of the Study of the Booke of Nature' with Boyle's earlier writings, from which natural philosophy is virtually absent (there is also the sudden change in the subject matter of the workdiaries to which I have already referred). Clericuzio does not really deny this, but he goes on to an ancillary claim, namely that 'Boyle's letters seem to disprove the view that Boyle's scientific career started in 1649 as a conversion.' To back this up, he cites various references, including the comments on the subjects studied by the Invisible College I have already quoted, and for which I had made allowance in my original claim.[65] In other words, there is no new evidence, simply a reassertion that these minor references *must* be more significant than I had allowed.

The point that needs to be reiterated, and that commentators like Clericuzio miss, is that the evidence for a seismic shift in Boyle's preoccupations from 1649 onwards derives from a comprehensive and chronologically sensitive survey of his extant manuscript remains as a whole. In my view, those who soak themselves in Boyle's profuse earlier writings and then read the works he started to produce

---

[63] See *Scrupulosity and Science*, pp. 34–5 (p. 73), and *Between God and Science*, p. 82.

[64] Clericuzio, 'Mercury in Mind', *Times Literary Supplement*, 4 June 2010. See also Clericuzio, 'The Many Facets of Boyle's Natural Philosophy', *Nuncius*, 23 (2008), 115–26, on pp. 118–20, and his review of Newman and Principe, *Alchemy Tried in the Fire*, in *Annals of Science*, 62 (2005), 406–8, on p. 407.

[65] Clericuzio, 'Mercury in Mind'. See also Clericuzio, 'Many Facets', p. 119, where he makes the same claim in virtually identical words except that he uses the date 1650 rather than 1649, which means that some of his supposed counter-evidence relates to 1649 – the very point at which I claim that Boyle's intensive scientific activity began. It is also worth noting that in ibid., p. 120, he claims that Boyle 'made experiments' with a camera obscura in Leiden in 1648, whereas in fact he was shown it as any tourist might have been (*Works*, vol. 2, p. 13), and that, though he cites Boyle's letter to Worsley about his attempts to construct a furnace, he ignores that to Lady Ranelagh about their failure (see above, n. 57).

in 1649 can only agree that a momentous change occurred at that point. Of course, this is perfectly compatible with the existence of piecemeal scientific references in Boyle's letters in the 1640s. But a minor enthusiasm for science as documented by the allusions of this kind to which I have referred is very different from the obsessive experimentalism that characterised the mature Boyle, and it is the roots of *this* that we are seeking. Indeed, as I pointed out in my 1995 article, many of Boyle's references in his 1646–47 letters to Hartlib and Worsley to his 'experiments' are rather scrappy and generalised, and do not suggest that he had much hands-on experience.[66] And why should he, at a time when he thought his predominant aim in life was lecturing his peers on the pursuit of virtue? Equally, if Clericuzio is right that Boyle was already doing science in the 1640s, why is it so absent from the profuse manuscript remains from this period that survive, in contrast to every subsequent period of his life? In other words, those who have taken an alternative point of view are guilty of tokenism, of grasping at straws because they have some preconceived wish to make Boyle a 'scientist' at the earliest possible stage in his life.

So I stick to my archivally based view of a 'Great Divide' in Boyle's life in 1649 – a claim which, incidentally, has been endorsed by various Boyle scholars and others.[67] Yet I should reiterate that, in many ways, it is a slightly surprising story, as Boyle is a slightly surprising figure. Indeed, I have come to feel more generally that it is the way he fails to fit into straightforward categories that is the key to understanding why Boyle was as significant as he was. I have emphasised this in trying to account for the success of the Boylean programme in science he put forward in the 1650s, the nature of which owed much to the sheer range of traditions on which he drew and the eclecticism he deployed in doing so.[68] One could also argue that Boyle was adept at sharing in traditions while capitalising on the degree of independence he retained. In the case of the Hartlib circle, this arguably enabled him to promote an effective, more secular version of the utilitarian and welfarist programme they had advocated in the aftermath of the Restoration. Here I am reminded of a fascinating letter from Worsley to Boyle, evidently dating from 1658 or 1659, that Tom Leng and I identified in the Hartlib Papers and which was included in *The Correspondence of Robert Boyle*. For what is interesting about this is the way in which it moves from a discussion of natural philosophy and medical reform to a highly charged account of the

---

[66]   See *Scrupulosity and Science*, pp. 21–2 (p. 63). It might be added that the same applies to the references to experiments in the Petty letter invoked in Clericuzio, 'Many Facets', p. 120.

[67]   Principe, 'Virtuous Romance', p. 392. See also ibid., p. 378; Newman and Principe, *Alchemy Tried in the Fire*, esp. pp. 10, 208–9; Dmitri Levitin, 'The Experimentalist as Humanist: Robert Boyle on the History of Philosophy', *Annals of Science*, 71 (2014), 149–82. For an intermediate view, see Leng, *Worsley*, p. 28 and pp. 25ff., passim.

[68]   See *Between God and Science*, ch. 7.

conflict between God and Satan in the world – the kind of juxtaposition that Boyle in his own writings eschewed, more or less consciously recognising that this was ill at ease with the ethos of English culture in the period of the Stuart Restoration.[69] His relations with another former Hartlibian protégé, Ralph Austen, and the advice he gave him to segregate his prescriptions on husbandry and theology when reissuing them in a Restoration context, reveals a similar strategy.[70] Equally, to return to what I said earlier about Boyle's relations with the Oxford group associated with Wilkins, his status as a layman in a predominantly clerical grouping gave him a power as a spokesman on which he was happy to capitalise in the years after 1660.[71] The unique and in some ways unexpected trajectory by which he first discovered science in a deep sense is another part of this story, and this is why it is important to vindicate it against views which have the tendency to homogenise Boyle and make him less interesting.

---

[69]   *Correspondence*, vol. 1, pp. 301–18. For a commentary, see Leng, *Worsley*, pp. 127–9. On Boyle's later adaptation of the Hartlibian ethos, see esp. *Scrupulosity and Science*, p. 212.

[70]   See *Correspondence*, vol. 2, p. 530; see also ibid., pp. 450–51.

[71]   *Between God and Science*, p. 176 and passim.

Chapter 3

Boyle and the Early Royal Society: A Reciprocal Exchange in the Making of Baconian Science[1]

The close relationship between Boyle and the Royal Society in its early years has been a commonplace ever since Boyle's own time. His name appears almost at the head of the list of the 12 men who inaugurated the society on 28 November 1660, second only to that of William, Viscount Brouncker, who was to become the society's first President. Equally important was Boyle's air-pump, the crucial piece of equipment which he had divulged earlier that year in his *New Experiments Physico-Mechanicall, Touching the Spring of the Air, and its Effects (Made for the most part, in a New Pneumatical Engine)*, and which became a kind of emblem of the society and its scientific programme in its formative period. Boyle presented a pump to the society in May 1661, and thereafter the demonstration and vindication of the findings of the 'pneumatical engine' became central to the society's early corporate life; as if symbolically, the modified version of the original device is depicted in the celebrated frontispiece to Thomas Sprat's promotional *History of the Royal Society* of 1667.[2]

Equally important was the way in which Boyle, more than anyone else, became the hero of the society's early protagonists. The *Philosophical Transactions* inaugurated by the first Secretary, Henry Oldenburg, are ceaselessly complimentary about Boyle, invariably there described in such formulae as 'that Noble Searcher of Nature'; Oldenburg devoted much space to flattering reviews of each of Boyle's books as they appeared, as well as to articles by him.[3] As Boyle's protégé, Oldenburg had an obvious interest in this, but the result of his efforts was undoubtedly to publicise Boyle as the principal exemplar of the Society's experimental policy. Boyle's iconic status for the society in its early years arguably

---

[1]   Previously published as 'Robert Boyle and the Early Royal Society: A Reciprocal Exchange in the Making of Baconian Science', *British Journal for the History of Science*, 40 (2007), 1–23.

[2]   Birch, *Royal Society*, vol. 1, pp. 3, 23 and passim; Thomas Sprat, *The History of the Royal Society of London*, 1667, ed. J.I. Cope and H.W. Jones (London, 1959), frontispiece; Steven Shapin and Simon Schaffer, *Leviathan and the Air-pump: Hobbes, Boyle and the Experimental Life* (Princeton, NJ, 1985), esp. ch. 2.

[3]   *Phil. Trans.*, 1 (1666), 153 and passim.

reached its climax in Joseph Glanvill's *Plus Ultra* (1668), another work which had a major influence whatever its ulterior motives. In defending the Royal Society against its critics, Glanvill devoted nearly two whole chapters to the work of 'the *Illustrious* Mr. *BOYLE*' as a means of illustrating the society's achievement:

> by a *single* Instance in one of their *Members*, who alone hath done enough to oblige all Mankind, and to erect an *eternal Monument* to his *Memory*. So that had this *great Person* lived in those days, when men *Godded* their *Benefactors*, he could not have miss'd one of the first places among their *deified Mortals*.[4]

This view of Boyle as the exemplar, and in many ways the inspiration, of the society's early activity has been echoed in recent scholarship, in which Boyle continues to occupy an almost emblematic role not dissimilar to that which he held for Oldenburg and Glanvill. In Paolo Rossi's *The Birth of Modern Science*, for instance, Boyle appears as 'one of the most influential members of the new institution', who was 'especially active' in its work, while a view of Boyle as the virtual archetype of the society is to be found in Steven Shapin's *A Social History of Truth*, in which we learn that Sprat's *History* 'was, in large part, the validation of pictures of the experimental philosopher and experimental social practices developed earlier by Robert Boyle'.[5]

Yet, considering the importance of Boyle for the Royal Society that such evaluations imply, it is perhaps slightly surprising to find that his links with the body were less continuous, and that he actually did less for it, than might have been expected. Thus, although Boyle acted as titular 'President' on one occasion before the society's incorporation in 1662, he thereafter failed ever to hold high office in the society, the most striking episode from this point of view occurring in 1680, when he was actually elected President but declined to serve.[6] Equally noteworthy is the fact that, in contrast to those stalwarts who regularly attended the meetings of the society through thick and thin, Boyle was distinctly haphazard in his attendance. This owed something to his peripatetic lifestyle for much of the 1660s, which meant that he was often out of London, but it is nevertheless significant, and if anything his attendance became more, rather

---

[4]   Joseph Glanvill, *Plus Ultra* (London, 1668), pp. 92–3 and chs 13–14, passim. For Glanvill's ulterior motives, see N.H. Steneck, '"The Ballad of Robert Crosse and Joseph Glanvill" and the Background to *Plus Ultra*', *British Journal for the History of Science*, 14 (1981), 59–74.

[5]   Paolo Rossi, *The Birth of Modern Science* (English trans., Oxford, 2001), pp. 199–200; Steven Shapin, *A Social History of Truth* (Chicago, IL, 1994), p. 185. See also ibid., pp. 129, 143, and ch. 4, passim, and Shapin and Schaffer, *Leviathan and the Air-pump*, p. 78 and ch. 2, passim.

[6]   Birch, *Royal Society*, vol. 1, p. 87, vol. 4, p. 58. For a discussion of the latter episode, see *Scrupulosity and Science*, pp. 64–8.

than less, erratic after he domiciled himself more permanently in the capital in 1668. If one studies the Royal Society's minutes, Boyle repeatedly disappears for lengthy periods. Thus he vanishes from September 1661 to March 1662 and from October to December 1662. He is absent again from May to August 1663, from August 1664 for the remainder of the year and then again from May 1665 to April 1666 (though in this case the society itself intermitted its activity for part of this time because of the Great Plague). Further gaps occurred from July 1667 to April 1668, from August 1668 to January 1669 and from July 1669 to March 1670 – despite the fact that in 1668 Boyle moved into Lady Ranelagh's house in Pall Mall.[7] Then, in June 1670, Boyle had a serious illness, and after this he attended less frequently, with a dozen or so appearances recorded between 1671 and 1674. Thereafter, he is only once recorded as being present, on 8 December 1680, during the debacle over his election to the Presidency.[8]

If all members had been as erratic in their attendance, the society could hardly have survived. Moreover, though Boyle is frequently recorded as speaking at meetings at which he was present, drawing attention to topics which he thought might interest the society or reporting on the progress of his experimental work, he often seems to have been pursuing an agenda of his own which did not necessarily overlap with that of the society. It is revealing that on one occasion when he was requested to 'give the society his thoughts' on the subject of fire, flame and heat, his response was 'that four or five years before he had made the consideration of this subject a part of his business, but did not know, whether his present studies of other matters would give him leave to review what he had then written'.[9] Overall, we are left with what might seem a distinctly one-sided relationship.

---

[7] For the principal gaps, see Birch, *Royal Society*, vol. 1, pp. 45–77, 115–67, 248–93, 455–511; vol. 2, pp. 50–83, 184–273, 312–40, 392–427. An additional point, drawn to my attention by Dan Purrington, is that when Boyle was on the society's Council in 1663–64, he often attended ordinary meetings of the society, but not the Council meetings that preceded them, suggesting a surprising lack of interest in the body as an institution: see *Between God and Science*, pp. 145 and 328 n. 5.

[8] Birch, *Royal Society*, vol. 2, pp. 500–501; vol. 3, pp. 21, 48, 50–51, 61, 63, 73–4, 76, 84, 88–9, 91–2, 114, 115–16, 131–2; vol. 4, p. 60. For ambiguous references which could imply Boyle's presence, but probably do not, see ibid., vol. 3, pp. 311, 510; vol. 4, pp. 252–3, 522. For Boyle's illness, see *Between God and Science*, pp. 174–5 and 334 n. 68.

[9] Birch, *Royal Society*, vol. 2, p. 8. There were, however, moments of real synergy between Boyle's interests and the society's, for instance in connection with transfusion experiments in 1666–67: see Harriet Knight and Michael Hunter, 'Robert Boyle's *Memoirs for the Natural History of Human Blood* (1684): Print, Manuscript and the Impact of Baconianism in Seventeenth-century Medical Science', *Medical History*, 51 (2007), 145–64, on pp. 146–7.

If this is the state of affairs from the point of view of the Royal Society, however, it is worth considering the matter more carefully from the point of view of Boyle. Was it simply a case of the Royal Society being the grateful and essentially passive recipient of the reflected glory of Boyle's role as a natural philosopher, happy to humour him when he deigned to take part in the society's activities? Or is there any evidence that, on the contrary, Boyle learned things from the Royal Society? For the biographer of Boyle, the great challenge is to ascertain when and why his ideas developed in the way they did. Obviously, conclusions on such issues will be based to a large extent on the profuse materials Boyle has left us, in the form of his voluminous published writings, his correspondence and his extensive papers. But herein lies a problem, because in these circumstances it is easy to succumb to 'navel-gazing' – to an exclusive focus on sources emanating from Boyle himself, failing to give sufficient attention to external factors unless they are actually referred to there. Yet this sometimes leaves explanatory *lacunae*, making one prone to underestimate the impact of significant stimuli that are only indirectly apparent from Boyle's own corpus. Indeed, in certain instances, Boyle seems to have been deliberately vague about what his sources actually were, hence providing a challenge to his interpreters to set the record straight. A case in point is his relationship with the German natural philosopher Daniel Sennert, where William Newman has illustrated how Boyle was less honest about the extent of his debt to the earlier author than many commentators have presumed.[10]

Arguably, a further case of such an underestimated stimulus may be represented by the influence on Boyle of the early Royal Society, in this case in relation to a key aspect of his natural philosophical method, his use of 'heads' and 'inquiries' as a means of eliciting and structuring data. This was a genre which, as we will see, went back to Francis Bacon. The most famous exemplar of it was Boyle's article 'General Heads for a *Natural History of a Country*, Great or small', published in *Philosophical Transactions* in 1666, with its sequel in the form of his 'Articles of Inquiries, touching *Mines*', but these belong to a larger corpus of such material that Boyle started to produce just at the time when these were published.[11] Thereafter, such 'heads' continued to play a significant role in his work for the rest of his career, as will be indicated below. But there is no obvious explanation from within Boyle's papers as to why this striking change in his working methods should have occurred at this particular time. If one looks at the fledgling Royal Society, on the other hand, one finds an enthusiasm for such 'heads of inquiries' which could almost be seen as a leitmotif of the

---

[10]   See W.R. Newman, 'The Alchemical Sources of Robert Boyle's Corpuscular Philosophy', *Annals of Science*, 53 (1996), 567–85, and the more general consideration in *Scrupulosity and Science*, pp. 144ff.

[11]   *Works*, vol. 5, pp. 508–11, 529–40. See further below.

society's activity in its earliest years, inspired by the Baconian imperative to data-collecting that was central to the society's rationale.[12] Here, it will be claimed that it was almost certainly due to the society's stimulus that Boyle came to see the methodological value of such 'heads'. Indeed, as we will see, it even appears that a committee of the society produced a set of 'General Heads' preceding Boyle's, though this is now lost.

* * *

To set the scene, it is necessary briefly to survey Boyle's writings on natural philosophy, in which there is a fairly clear pattern of development from the point when he first discovered science around 1650. There is thus continuity from the largely apologetic writings that comprise his earliest extant natural philosophical works to such published books as *The Usefulness of Natural Philosophy*, largely compiled in the late 1650s and published in 1663 and 1671.[13] Then, in his 'Oxford' period from 1656 onwards, Boyle developed an influential programme to which the classic 'Essay on Nitre' which formed the core of his *Certain Physiological Essays* (1661) was central. This led to a whole series of discursive essays in which he sought to use experimental evidence to vindicate the mechanical hypothesis. Some of these were published in the 1660s, including his famous *Origin of Forms and Qualities*; others came out in extended form at a much later date.[14] All these writings take the form of ruminative essays, in which experimental data was adduced, but in a far from systematic way, and this was a style of writing that Boyle continued to use in much of his scientific work for the rest of his life. Even *Spring of the Air*, though presenting a series of numbered experiments, takes a comparably discursive form.

In 1665, however, Boyle published a book of a rather different kind, his *New Experiments and Observations Touching Cold*. In contrast to the writings already noted, the experiments that are reported in the main part of this massive work are divided up into thematic 'Titles', dealing successively and systematically

---

[12]   On the early Royal Society's Baconianism, see Michael Hunter, *Establishing the New Science: The Experience of the Royal Society* (Woodbridge, 1989), esp. chs 1, 3 and 6; W.T. Lynch, *Solomon's Child: Method in the Early Royal Society of London* (Stanford, CA, 2001), and W.T. Lynch, 'A Society of Baconians? The Collective Development of Bacon's Method in the Royal Society of London', in J.R. Solomon and C.G. Martin (eds), *Francis Bacon and the Refiguring of Early Modern Thought* (Aldershot, 2005), pp. 173–202 (Lynch's views and my own are more compatible than he implies). On the use of inquiries, see Daniel Carey, 'Compiling Nature's History: Travellers and Travel Narratives in the Early Royal Society', *Annals of Science*, 54 (1997), 269–92, and below.

[13]   *Works*, vol. 13, pp. xxxvii–xlii and passim, and vol. 3, pp. xix–xxiv, 189ff.; vol. 6, pp. li–liv, 389ff..

[14]   Ibid., vol. 1, pp. xxxiii–xxxiv, esp. p. xxxiiin.

with experiments on such topics as 'Bodies capable of Freezing others', 'Bodies disposed [or indisposed] to be Frozen', 'the degrees of Cold in several Bodies', 'the Tendency of Cold, upwards or downwards' and so on.[15] This was the first of Boyle's books to be promoted in the newly founded *Philosophical Transactions*, and, if anything, the synopsis of the work published there brought out its structure all the more clearly in summarising Boyle's main findings under its various heads.[16] In his preface, Boyle refers to 'the Scheme of heads of Inquiry, that I drew up to give myself a general Prospect of the subject I was to handle'; this provided him with a 'method' for his material which could at the same time accommodate 'future discoveries and hints' that he or others might add. As Boyle put it: 'being unwilling to huddle my Experiments confusedly together, I thought it an expedient ... to draw up a company of comprehensive Titles, under which might commodiously be rang'd most of the Particulars I had observ'd'.[17] In this work, for the first time, Boyle imposed a clear and orderly structure on his profuse experimental investigations of a whole class of phenomena concerning the natural world.

What was his source? The clue is provided by the title page, which has on it a quotation from Francis Bacon, though, oddly enough, his inspiration was not otherwise very explicitly acknowledged in *Cold* itself.[18] It was in a later work with a comparable structure, Boyle's *Memoirs for the Natural History of Human Blood* (1684), that he clarified the origin of his new format. There, he explained of the various 'heads' he used to subdivide his subject: 'I thought fit to comprize all these sorts of particular Topicks, or Articles of Inquiry (to use our illustrious *Verulam*'s phrase) under the general and comprehensive name of *Titles*.'[19] Bacon had indeed opened the two principal specimens of the natural and experimental histories which he saw in his *Instauratio magna* as 'the very foundation of our work' with just such lists of 'Topica Particularia': these were his 'History of Winds' and his 'History of Life and Death', published in his *Historia naturalis et experimentalis ad condendam philosophiam* of 1622. As he explained at the outset,

---

[15]   Ibid., vol. 4, pp. 210–11, 226–7 and 213ff., passim. It is perhaps worth noting that the prolegomena concerning the thermometers and other equipment used and the authors cited, and the appendages discussing alternatives theories concerning cold, could be seen to exemplify the proper structure for a natural history expounded in Boyle's 1666 'Designe for Natural History': see below.

[16]   *Phil. Trans.*, 1 (1666), 8–9, 46–52, and esp. the latter, though it peters out at title 13, due to want of space.

[17]   *Works*, vol. 4, p. 210.

[18]   Ibid., p. 203. The quotation is from Bacon, *Novum organon*, ii, 10: see Bacon, *The Instauratio magna Part II: Novum organum*, ed. Graham Rees with Maria Wakely (Oxford, 2004), pp. 214–15.

[19]   *Works*, vol. 10, p. 9. The discussion of his sources here supercedes the note given there.

in 'the Rule of the present History': 'In the particular titles, after a preamble or preface, I at once lay down particular topics or articles of inquiry both as light to the present and stimulus to future inquiry.'[20] Though these lists and their role in Bacon's thought have not been much studied by Bacon scholars, in both cases they do indeed seem to have played a crucial role in structuring the work, being cited at the start of each section, and evidently providing an agenda both for the accumulation of relevant information and for the content of the text. A handful of other such lists survive, dealing with such topics as magnetism, light, metals and minerals, the transmutation of bodies and the commixture of liquors.[21] In addition, the idea of compiling such lists was divulged by Bacon in his influential *Parasceve* or 'Preparative for a Natural and Experimental History', where he listed over a hundred such topics, explaining how, 'as soon as we have the leisure for the task, we plan to give detailed instructions by putting the questions that most need to be investigated and written up in each history because they help to fulfil our purpose, like certain particular *Topics*'.[22] Bacon thus provided a clear model for Boyle.

---

[20]   Bacon, *The Instauratio magna Part III: Historia naturalis et experimentalis: Histora ventorum and Historia vitæ et mortis*, ed. Graham Rees with Maria Wakely (Oxford, 2007), pp. 12–15, 18–29, 150–55. See also *Instauratio magna Part II: Novum organum*, pp. 214–15. The issue of Bacon's possible sources cannot be addressed here, but among precedents it is worth noting (1) the Aristotelian 'problem literature' which Bacon saw as worthy of emulation in his *De augmentis scientarum* (Brian Lawn, *The Salernitan Questions* (Oxford, 1963), p. 141 and passim); for the scholastic background to the genre, see also Peter Anstey, 'The Methodological Origins of Newton's Queries', *Studies in History and Philosophy of Science*, 35 (2004), 247–69, esp. pp. 248–9, 252–3; (2) the genre of ecclesiastical visitation articles (see W.H. Frere (ed.), *Visitation Articles and Injunctions of the Period of the Reformation*, (3 vols [Alcuin Club, vols 14–16], London, 1910), and (3) the questionnaires addressed to the Spanish possessions in the New World in the late sixteenth century (see, for example, H.F. Cline, 'The *Relaciones Geográficas* of the Spanish Indies, 1577–86', *Hispanic American Historical Review*, 44 (1964), 341–74). The possible influence of Bacon's legal background has not been explored.

[21]   These were published in the posthumous *Scripta*, edited by Jan Gruter in 1653, in the *Opuscula*, edited by William Rawley in 1658, and in *Baconiana*, edited by Thomas Tenison in 1679 (the latter too late, incidentally, to have influenced Boyle): see Francis Bacon, *The Instauratio magna: Last Writings*, ed. Graham Rees (Oxford, 2000), pp. 237–57, and *Works*, ed. James Spedding, R.L. Ellis and D.D. Heath (14 vols, London, 1857–74), vol. 3, pp. 799–826.

[22]   Quoted from Bacon, *The New Organon*, ed. Lisa Jardine and Michael Silverthorne (Cambridge, 2000), p. 232. See also ibid., pp. 233–8, and Bacon, *The Instauratio magna Part II: Novum organum*, pp. 472–3, 474–85. Bacon also proposed the use of 'Topica quædam inductiva, sive Articulos ad interrogandum' in his *Descriptio globi intellectualis*: see Bacon, *Philosophical Studies c. 1611–c.1619*, ed. Graham Rees (Oxford, 1996), pp. 114–15.

Yet in his profuse writings prior to *Cold*, Boyle had failed to follow Bacon's example – a fact which has been overlooked partly because it is so commonplace that Boyle was a Baconian that the actual trajectory of his Baconianism has been little studied.[23] In fact, Bacon does not play a particularly prominent role in Boyle's early natural philosophical writings. In Boyle's earliest discussions of scientific topics, Bacon is alluded to in a generalised way as a 'Free Philosopher' along with such authors as Telesio, Campanella, Sennert and Gassendi.[24] Even in *Usefulness*, Bacon is only occasionally cited, usually on quite specific points.[25] In *Certain Physiological Essays*, Boyle was ambivalent. In the 'Proemial Essay', he protested how he had refrained from reading Bacon's *Novum organum*, along with the natural philosophical writings of authors like Descartes and Gassendi, 'that I might not be prepossess'd with any Theory or Principles till I had spent some time in trying what Things themselves would incline me to think' – a slightly gnomic statement whose significance I have discussed elsewhere.[26] Later in the same essay, however, he expressed the ambition to continue Bacon's famous *Sylva sylvarum* in the form of a collection of 'Promiscuous Experiments', which is significant, if representing a rather unsophisticated ambition by comparison with the structured format used in *Cold*.[27] A similar transitional stage is represented by Boyle's *Experiments and Considerations touching Colours*, published in 1664. After opening with a digressive account of the phenomena in hand addressed, like those in *Certain Physiological Essays* and *Usefulness*, to Pyrophilus (the name given by Boyle to his nephew Richard Jones, later First Earl of Ranelagh), the rest of the book is quite different in format. It comprises a series of 15 'Experiments in Consort, Touching Whiteness & Blackness', and then a 'Third Part. Containing Promiscuous Experiments About Colours' numbering 50 – in other words, the equivalent of half of one of the 'centuries' of experiments which *Sylva sylvarum* comprised, and hence possibly exemplifying the ambition to continue that work referred to in *Certain Physiological Essays*.[28]

---

[23] For an account of Boyle's Baconianism, see Rose-Marie Sargent, *The Diffident Naturalist: Robert Boyle and the Philosophy of Experiment* (Chicago, IL, 1995): however, she does not comment on the evolution of Boyle's Baconianism as outlined here, and she mentions the 'heads' and 'inquiries' only in passing on pp. 175 and 297 n. 70.

[24] *Works*, vol. 13, pp. 190, 197. Cf. the citation of *De augmentis scientarum* in ibid., p. 157.

[25] Ibid., vol. 3, pp. 229, 271, 349, 364, 433–4, 477, 536; vol. 6, pp. 402, 409, 432–3; vol. 13, pp. 351, 353. Note that in the last text, which is of fairly late date, Boyle says of the *Novum organum* that he was 'not then vers'd' in it when he initially wrote this essay.

[26] *Works*, vol. 2, pp. 12–13; *Scrupulosity and Science*, pp. 145–6.

[27] *Works*, vol. 2, pp. 17–18.

[28] Ibid., vol. 4, pp. 3ff., passim. *Colours* had the same motto from Bacon on its title page as did *Cold*.

With *Cold*, however, the Baconian structure is much more overt, and this is also true of at least three other works by Boyle, all of them, significantly, with roots going back to the 1660s, though they were not published till much later. These are *Memoirs for the Natural History of Human Blood*, published in 1684, *Short Memoirs for the Natural Experimental History of Mineral Waters*, published in 1685, and *The General History of the Air*, posthumously published by John Locke in 1692.[29] The earliest version of the set of 'Titles' used in *Human Blood* survives in a notebook of Locke's dating from the 1660s, while both of the other works also seem to have their origins in that decade.[30] All of these present lists of 'topics' or (to use Boyle's preferred word) 'titles', while the remainder of their content comprises a collection of data categorised under one or other of the headings there laid out.

Equally significant is the fact that Boyle's papers contain a series of nearly twenty lists of such 'heads', though this is less well known because until recently they survived only in manuscript. These range in their descriptions from the almost directly Baconian 'Titles or Topica Particularia about the Natural History of Water' to the more general 'Enquirys and Experiments about Electricall Bodys'; others simply give a list of headings without an overall title.[31] There is some variety in the range and scale of such items. Some, such as 'Titles and Articles of Inquiry in Order to A Natural History of the Sea', effectively sketch out the contents of a planned book, in this case running systematically through every aspect of sea water, its nature and content, before dealing with storms, currents and other related phenomena. Others, on the other hand, are more specific lists of trials to be made, as with 'Experiments to be made in seald Receivers', which simply itemises a series of substances to be tested thus.

---

[29] Ibid., vol. 10, pp. 3ff., esp. 12–13, 15–16, 40–1; 205ff., esp. 220–4, 247–9; vol. 12, pp. 3ff., esp. 7–9. See also the recently-published 'Memoirs for the Natural History of Tin', ibid., vol. 14, pp. 133ff. In contrast to the rather scrappy nature of certain of these works, it is perhaps also worth noting here that the closest parallel to *Cold* among Boyle's later writings is his 'New Pneumatical Experiments about Respiration', published in *Phil. Trans.* in 1670 (*Works*, vol. 6, pp. 213–57), which has 20 'Titles', with a varying number of experiments under each; however, it does not have a list of the titles at the start.

[30] Bodleian Library, Oxford, MS Locke f. 19, pp. 272–3, 302–3, the former (only) reproduced in Kenneth Dewhurst, 'Locke's Contribution to Boyle's Researches on Air and on Human Blood', *NRRS*, 17 (1962), 198–206, pl. 12. For the text of this and the various revised versions of the list, see Michael Hunter and Harriet Knight (eds), *Unpublished Material Relating to Robert Boyle's 'Memoirs for the Natural History of Human Blood'*, Robert Boyle Project Occasional Papers No. 2 (London, 2005). For the link of this and other writings to the 1660s, see Boyle, *Works*, vol. 10, pp. xi–xii, xxix–xxx; vol. 12, pp. xi–xiv.

[31] See Michael Hunter (ed.), *Robert Boyle's 'Heads' and 'Inquiries'*, Robert Boyle Project Occasional Papers No. 1 (London, 2005). This gives a complete text of the documents with a commentary. For a listing, see below, n. 33.

Either way, the document outlined the leading features of a phenomenon that seemed worthy of investigation, more often than not by experimental inquiry. An untitled series on elasticity, for instance, opens with the following:

> What Bodys are Naturally endowd with elasticity.
>
> What Bodys Naturally want Springs.
>
> What Bodys there are that have Springs under some Dimensions, & not under others, & but what measures as to length & thicknesse they appear to have, & not to have a Spring.
>
> What Bodys not Naturally or always elasticall are capable of being made soe.
>
> What Bodys Naturally Elasticall may be depriv'd of their Spring (to this belongs the Glasse of Lead & other Minerales per se, & the reductions of that Minerall).
>
> By what operations & meanes Elasticity may be introduc'd into Bodys, as fusion hammering, wire drawing &c.
>
> By what Operations & meanes the Elasticity of a Body may be destroyd, as nealing, melting &c.
>
> What are the cheif & most usuall Concomitants of Elasticitie, & of the absence or losse of it.[32]

What is particularly significant is that these lists survive to a disproportionate extent in the handwriting of amanuenses employed by Boyle in the 1660s. This is important because handwriting clues of this kind have proved crucial in identifying 'strata' within the Boyle Papers, and it is unusual to find such a high preponderance of 1660s hands in a group of material, as against handwritings from earlier or later in Boyle's career.[33] Revealingly, none at all are of earlier

---

[32] BP 10, fol. 132. This comprises the first third of the document only. For the remainder, see Hunter, *Boyle's 'Heads' and 'Inquiries'*, pp. 4–6.

[33] See *Works*, vol. 1, pp. c–cii, and *Boyle Papers*, ch. 1. The following items are in 1660s hands: BP 10, fols 57 (magnetical trials; hand G; Hunter, *Boyle's 'Heads' and 'Inquiries'*, pp. 2–3), 118 (experiments in sealed receivers; hand E; *Boyle's 'Heads'*, pp. 3–4), 132 (elasticity; hand F; *Boyle's 'Heads'*, pp. 4–6), 133v–134 (tastes; hand F; *Boyle's 'Heads'*, pp. 6–7), 135v–136 (odours; hand F; *Boyle's 'Heads'*, pp. 7–8); BP 18, fols 129–30 (anatomical experiments; hand F; *Boyle's 'Heads'*, pp. 9–12); BP 22, pp. 197–200, continued in BP 38, fol. 120 (electrical bodies; hand F; *Boyle's 'Heads'*, pp. 12–15); BP 25, pp. 392–3 (shining wood; hands E and F; *Boyle's 'Heads'*, pp. 16–17); BP 26, fols 75–6 (lime; hand F; *Boyle's 'Heads'*, pp. 24–6), 90 (insects and 'Spontearta'; hand F; *Boyle's 'Heads'*, pp. 26–7); Boyle Letters 3, fols 33–4 (Greatrakes; hand F; *Boyle's 'Heads'*, pp. 31–2). In addition, BP 10, fol. 5 (volatile salts; *Boyle's 'Heads'*, pp. 1–2), is in the hand of Frederic Slare, who started to work for Boyle *c.* 1670. The following are in the hand of Boyle's amanuensis Robin Bacon, who worked for Boyle from the mid-1670s onwards: BP 25, pp. 264–5, and BP 26, fols 49–50 (natural history of water; two versions; *Boyle's 'Heads'*, pp. 17–19); BP 26, fols 51v–52 (natural history of the sea; *Boyle's 'Heads'*, pp. 19–23); BP 26, fols 62, 70

date, and though a minority are in later hands, either because composed or recopied later, it is worth observing that the 1660s items are often clearly actual 'composition copies' – as shown by the fact that they contain various deletions and additions made in the hand of the amanuensis who wrote them in the first place – whereas this is true to a much lesser extent of the later ones.[34]

The link of such lists of titles with the 1660s is further documented by various references to these texts in Boyle's profuse correspondence with Henry Oldenburg in these years.[35] He also wrote discursively about such lists and their significance for natural history in his 'Designe about Natural History', which takes the form of a letter to Oldenburg dated 13 June 1666, from which he later extracted and extrapolated in the disquisition that opened *Human Blood*.[36] This text is notable for its sophisticated treatment of natural history as a whole, both in its openness to the assessment of contrasting hypotheses and its preparedness to emend Bacon's categories when appropriate, reflecting the fact that Boyle was by this time giving careful thought to Bacon's strictures on the writing of natural history and elaborating on these. He proposed a comprehensive, if provisional, categorisation of the subject matter of natural history into various 'principal Parts', and the idea was evidently to subdivide these in a manner comparable to Bacon's *Parasceve*, though, if Boyle executed this, it has not survived other than in the form of the examples he gave in the 'Designe' to illustrate each of his general categories, and the piecemeal sets of 'titles' that are extant.[37]

---

and BP 36, fols 98–9 (light and luminosity; four copies; *Boyle's 'Heads'*, pp. 23–4); BP 36, fol. 80 (copper; *Boyle's 'Heads'*, pp. 30–31); British Library Sloane MS 2502, fols 1v–2 (diseases; *Boyle's 'Heads'*, pp. 33–4). Documents in other hands: BP 27, pp. 331–2 (gems; hand: Oldenburg; *Boyle's 'Heads'*, pp. 28–30); Bodleian Library MS Locke c. 42, fols 266–7 (flame and fire; hand: Brownover; *Boyle's 'Heads'*, pp. 34–6); Bodleian MS Lister 34, fol. 35 (damps: hand: Oldenburg; *Boyle's 'Heads'*, pp. 36–7).

[34] See Hunter, *Boyle's 'Heads' and 'Inquiries'*, passim. For a later text showing evidence of revision, see the two versions of the titles for the natural history of water, BP 25, pp. 264–5, and BP 26, fols 49–50 (*Boyle's 'Heads'*, pp. 17–19).

[35] See *Correspondence*, vol. 3, p. 80 (quoted in text below, p. 73: probably a reference to BP 26, fol. 90; Hunter, *Boyle's 'Heads' and 'Inquiries'*, pp. 26–7), 251–2 (this refers to the queries published in *Phil. Trans.*, but BP 26, fol. 51v–52 (*Boyle's 'Heads'*, pp. 19–23), may be linked to these), 373, 376, 380 (the former evidently referring to BP 10, fol. 118 (*Boyle's 'Heads'*, pp. 3–4), as well as BP 25, pp. 392–3 (*Boyle's 'Heads'*, pp. 16–17), on which, see further below). For references to the inquiries published in *Philosophical Transactions*, see below, nn. 69–70.

[36] *Correspondence*, vol. 3, pp. 170–75, and Michael Hunter and Peter Anstey (eds), *The Text of Robert Boyle's 'Designe about Natural History'*, Robert Boyle Project Occasional Papers No. 3 (London, 2008); *Works*, vol. 10, pp. 9–12.

[37] See Peter Anstey and Michael Hunter, 'Robert Boyle's "Designe about Natural History"', *Early Science and Medicine*, 13 (2008), 83–126. See also Peter Anstey, 'Locke, Bacon and Natural History', *Early Science and Medicine*, 7 (2002), 65–92.

As is apparent from their proliferation among his literary remains at this point, Boyle clearly found such sets of 'heads' useful for various reasons. Most obviously, for him as for Bacon they acted as a means of setting himself an agenda and clarifying and structuring his ideas on his chosen subject. In *New Experiments Touching Cold*, as we have seen, he saw such 'titles' as providing 'a general Prospect of the subject I was to handle', while, in justifying the compilation of such lists of '*Topica particularia* or Articles of Inquiry' in a document that almost certainly belongs to his 'Designe about Natural History', he explained how 'tis highly useful for the discovery of the nature of a Body to consider how many wayes it may be examin'd, or (if you will) how many distinct *Phænomena* and representations of itself, it may be made to exhibit'.[38] An example of the way in which this might relate to his experimental work is provided by one of his manuscript lists, in this case headed 'Concerning shining Wood', with columns headed 'Observations to be made' and 'Tryalls to be made'.[39] This is apparently the document referred to in Boyle's account of his experiments on luminescent bodies published in *Philosophical Transactions* in 1668, where he explains how he had formerly collected observations on this topic, and 'had also proposed several Trials about them, to be made when I should have opportunity, and requisite Instruments to put them in practice'. In fact, as he goes on to explain, the experiment he tried was 'but one' of those 'in my List' (where it does indeed appear), but Boyle found it so engrossing that he went on to make a whole series of ancillary experiments, adding a series of reflections 'About the *Resemblances* and *Differences* between a *Burning Coal* and *Shining Wood*', which well illustrate his experimental virtuosity. On the other hand, this still left him with the original list as set out in his manuscript, from which he simply deleted this particular item, renumbering the remainder as an agenda for further investigation.[40]

Equally important, such lists could be used as a means of eliciting information relevant to a topic from others, and for classifying such data once it had been collected. Indeed, the extent to which they combined the provision of a structure for the study of a topic with an agenda of enquiry is apparent both from the Baconian precedent – in which 'topica particularia' were routinely combined with 'articles of inquiry' – and from Boyle's own adaptation of it. As he explained in *Human Blood*, his 'Titles' combined '*Queries* more properly So called, *Propositions* either Affirmative or Negative, and other Heads of Natural History'.[41] Indeed, in his descriptions of these documents, 'titles' or 'heads' overlapped seamlessly with 'queries' or 'inquiries' – often without this making

---

[38]   *Works*, vol. 4, p. 210; BP 9, fol. 71, printed in Hunter and Anstey, *The Text of Boyle's 'Designe'*, p. 6. It has 'particularis' for 'particularia' – an obvious mistranscription.

[39]   BP 25, pp. 392–3; *Boyle's 'Heads'*, pp. 16–17.

[40]   *Works*, vol. 6, pp. 1–25, esp. pp. 5, 20; BP 25, p. 393 (*Boyle's 'Heads'*, p. 17).

[41]   *Works*, vol. 10, p. 9.

much difference to the mode of presentation of the document in question, in which entries starting 'Whether' were interchangeable with others starting 'Of'. Thus, for instance, an untitled series relating to copper oscillates between entries such as 'Of the Fusiblenes of Copper' and 'Whether Copper be an Homogeneous Body?'[42]

That in Boyle's case he saw his 'titles' as an appropriate tool for eliciting information from others is shown most clearly by the fact that his 'Titles of the (Naturall and Experimental) History of the Air' – a typical list of topics mainly set out in a descriptive rather than an interrogatory tone – was apparently printed (though no copy survives); there is also evidence from Boyle's *Correspondence* that it was circulated, eliciting information on various of the matters it covered.[43] Moreover, the overlap between 'titles' and 'queries' is illustrated perhaps most clearly by Boyle's best-known compositions of this kind, published, as already noted, in *Philosophical Transactions* in 1666 – his 'General Heads for a *Natural History of a Country*, Great or small' and his 'Articles of Inquiries touching *Mines*', which are then subdivided by 'Titles'.[44] This overlap between 'heads' and 'inquiries' is slightly strange to us, since we are prone to expect documents intended to elicit information to have an explicitly interrogatory tone, yet one should note how these grew out of the structured lists of topics with which, as we have seen here, they overlapped.

In addition to using these lists to interrogate others, Boyle also employed them to categorise data so accumulated. This has been revealed by Harriet Knight in her recent PhD thesis on Boyle's methods of intellectual organisation, in which she has analysed the content of a massive compilation entitled 'Promiscuous Experiments, Observations & Notes', the most extensive of the series of workdiaries that Boyle drew up throughout his intellectual career, in which he recorded both his own experimental and observational findings and information divulged him by others. She has shown that the entries have been endorsed according a scheme which is clearly that of Boyle's 'General Heads', with material being allocated a number according to whether it was best placed under the heading 'Air', 'Water', 'Earth', 'Mines', 'Plants', 'Animals', 'Inhabitants' or human productions.[45] A similar rationale is expressed by a paper entitled 'Advertisements', which seems to have been intended to accompany such a collection of data and to clarify its relationship to the 'Naturall Historys' to which

---

[42]   BP 36, fol. 80; *Boyle's 'Heads'*, pp. 30–31.

[43]   See *Correspondence*, vol. 6, pp. 123, 132–3, 151–2, 180, 217–23. The set of 'titles' to which John Clayton was replying in the last reference differed slightly in its numeration from that in *Works*, vol. 12, pp. xii, xxiii–xxiv. On the printing of the 'Titles', see ibid., xi–xii, 5.

[44]   *Works.*, vol. 5, pp. 508–11, 529–40. For their date, see below.

[45]   Harriet Knight, 'Organising Natural Knowledge in the Seventeenth Century: the Works of Robert Boyle' (PhD thesis, Birkbeck, University of London, 2003), pp. 110–13. The workdiary in question is no. 21: see http://www.livesandletters.ac.uk/wd/index.html.

Boyle aspired. In it, he states how 'the particulars that are here huddled together in the casuall order wherein they occurd to me, are to be rang'd according to the Intimations of the *Topica particularia* or Articles of Enquiry about each of these Historys', the histories in question being enumerated as 'Fermentation, Putrefaction, water, air, flame &c'. – in other words, categories of the sort that might have been predicted from his 'Designe about Natural History'.[46]

Hence the lists of 'heads' that Boyle evidently started to compile at this time almost immediately began to serve a crucial role in his intellectual agenda. But what had stimulated their inception? Within Boyle's manuscripts, there are no real clues to help answer this question – documents of this type simply appear for the first time in the handwriting of amanuenses whom he employed in the 1660s. But here we need to return to the Royal Society, because it is almost certainly not coincidental it was in the society's early years that Boyle first developed this enthusiasm for using such 'heads' – not least since, as we shall see, the institution was from its inception much preoccupied by the preparation of elaborate lists of queries, as the most effective way of organising the data-collecting that was central to its ambitions. First, however, it is worth returning to *New Experiments and Observations Touching Cold* (1665), the first work by Boyle to be organised under 'titles', for this was very much a 'Royal Society' book, more so than almost any other publication by Boyle.[47] Though the material on which it was based seems to have been collected over a longer period, it was prepared for publication in the society's early years, and it was evidently at this stage that the book's structure originated. In his preface, Boyle specifically linked the inception of the work to '*The Command of the Royal Society*', reiterating in a letter to the President, Lord Brouncker, which precedes the main text, how its collection of experimental data was drawn up at the society's behest. He also apparently circulated copies of certain sections of the work to the Fellows of the society prior to publication, including the list of 'heads' under which the experiments were organised.[48]

Also worth considering here is the case of Boyle's former employee, Robert Hooke, who had become the first Curator of Experiments to the Royal Society in November 1662. One of Hooke's earliest activities in this role, in February 1663, was to present the society with an elaborate 'scheme of inquiries concerning the air', which formed the basis of a systematic programme of experiments that he was

---

[46] BP 10, 138, published in *Boyle Papers*, pp. 183–4. This, too, has 'particularis' for 'particularia' (see above, n. 38).

[47] The main potential rival is *Hydrostatical Paradoxes* (1666).

[48] *Works*, vol. 4, pp. 206, 210, 263–4. See also ibid., pp. xvii–xxi. It is odd that there is no reference in the minutes published by Birch to the explicit instruction to produce the work to which he there appears to refer. However, for the copies of the heads etc., see Birch, *Royal Society*, vol. 2, pp. 2, 5.

subsequently to pursue at the society's meetings.[49] Thereafter, the compilation of documents of this kind became a major preoccupation for Hooke, as seen particularly in his 'General Scheme, or Idea Of the Present State of Natural Philosophy, And How its Defects may be Remedied', a remarkable exposition of scientific method on essentially Baconian principles which dates from the late 1660s although it was not published until after Hooke's death. This was almost certainly based on presentations made to the Royal Society, and in it Hooke gave a series of such lists of queries.[50] Indeed, Hooke was implicitly as systematic in this regard as Bacon himself, seeing the examples that he gave (which related to astronomical phenomena and to various aspects of the air) as merely a 'few Instances' which 'may serve for a Specimen of what I mean by the Method of propounding Queries on any Subject, to be examined by accurate Observations and Tryals, before the Writing a Natural History of it'.[51]

Here it is significant that we specifically know that the 'scheme of inquiries concerning the air' that Hooke produced in February 1663 was compiled on the orders of the Fellows of the Royal Society as a collective body.[52] Without belittling Hooke in any way, this implies that, in drawing up this pioneering list, he was responding to a corporate initiative on the part of the society, which evidently had an effect on him comparable to that which it seems to have had on Boyle. We arguably tend to underestimate the extraordinary collective energy of the Royal Society as a body in its formative years, which transcended the role of individuals and in which the whole was truly more than the sum of its parts. Moreover, within it, whereas we are prone to be preoccupied by individual Fellows who are retrospectively famous, such as Hooke and Boyle, an equally important role was apparently played by less well-known figures, not least in formulating the society's programme. A case in point is Sir Robert Moray, whose memorandum on the society's 'business & designe' has often been seen to epitomise the institution's aims in its early years, and whose enthusiasm for structured data-collection is seen, for instance, in his '*Patternes of the Tables* proposed to be made for *Observing of Tides*' published in *Philosophical*

---

[49]   Ibid., vol. 1, pp. 202–4 and passim. See also Hunter, *Establishing the New Science*, p. 23.

[50]   Robert Hooke, *Posthumous Works*, ed. R. Waller (London, 1705), pp. 1–70, esp. pp. 21–33. For an associated item, referred to there, see Hooke's 'Cometa', in his *Lectiones Cutlerianae* (London, 1679), pp. 7–8, reprinted in R.T. Gunther (ed.), *Early Science, in Oxford* (14 vols, Oxford, 1923–45), vol. 8, pp. 223–4. For the link to the Royal Society Cutlerian lectures, see Hunter, *Establishing the New Science*, pp. 301–3. On these 'queries', see Anstey, 'Methodological Origins', pp. 254–6.

[51]   Hooke, *Posthumous Works*, p. 33.

[52]   Birch, *Royal Society*, vol. 1, pp. 198, 202.

Transactions in 1666.[53] The first President, Lord Brouncker, may have played a similar role. Boyle's dedication to him of *Cold* could reflect his particular significance in stimulating the way in which that work was presented, while, previous to this, it is interesting that he and Boyle had collaborated on what seems to be the earliest document of the kind under consideration here: this was 'some questions, in order to the tryal of the quicksilver experiment upon Teneriffe' which a group of five Fellows, including Brouncker and Boyle, was asked to compile at the first full meeting (on 5 December) of the society after its inauguration on 28 November 1660, and which Boyle and Brouncker produced at the meeting on 2 January 1661. This was the very first item to be entered in the society's Register Book, thus almost symbolising the association of such texts with the nascent society.[54] It is perhaps significant that neither Brouncker nor Moray had been active in the 'Oxford group' with which Boyle had been associated in the late 1650s, hence helping to explain the distinctive character of the Royal Society's apparent institutional influence on him and others at this point.[55]

Above all, this corporate initiative is seen in an enthusiasm for the preparation of sets of 'inquiries' for those travelling to exotic places, and, since this forms the background to arguably the most influential of all the compilations of this kind that emanated from the Royal Society in its early years, in the form of Boyle's 'General Heads for a *Natural History of a Country*, Great or small', this matter deserves detailed consideration. On 6 February 1661, a committee of 16 Fellows was appointed 'for considering of proper questions to be inquired of in the remotest parts of the world', the list being led by Brouncker and Moray, and again including Boyle along with a number of Fellows reflecting the metropolitan, governmental input into the society, including Thomas Povey, the eminent colonialist; William Petty, projector and courtier; Lawrence Rooke, Gresham Professor of Geometry; Henry Oldenburg, the future Secretary of the society; the virtuosi John Evelyn and Thomas Henshaw, and the merchant Daniel Colwall.[56] Equally interesting is the fact that on 4 March 1661 – still within

[53] For the attribution of the former to Moray, see Michael Hunter, *Science and the Shape of Orthodoxy* (Woodbridge, 1995), pp. 172–4; for a commentary on it, see Lynch, 'Society of Baconians?', pp. 187–8 and passim. For the latter, see *Phil. Trans.*, 1 (1666), 311–13.

[54] Birch, *Royal Society*, vol. 1, pp. 5 (the other Fellows asked were Moray, William Petty and Christopher Wren), 8–10; Royal Society Copy Register Book (hereafter RBC), vol. 1, pp. 1–3. For Brouncker's pneumatic experiments, see *Works*, vol. 3, p. 62.

[55] Brouncker was in epistolary communication with a key Oxford figure, John Wallis (see J. Wallis (ed.), *Commercium epistolicum* (London, 1658), passim), but there is no evidence that he had any contact with Boyle at that point.

[56] Birch, *Royal Society*, vol. 1, p. 15. Cf. ibid., p. 17. The other members were John Wilkins; William Coventry, courtier and politician; the doctors Jonathan Goddard, Daniel Whistler and Timothy Clarke; and John Austen, a somewhat shadowy figure. It is perhaps

weeks of the society's inauguration, and representing its earliest documented public initiative – we find Thomas Povey sending a correspondent in Virginia, Edward Digges, a set of 'Enquiryes concerning those severall kind of things which are reported to be in Virginia & the Bermudas, not found in England'.[57] He explained how these emanated from the society, described as:

> certaine Noble and Ingenuous Persons, who by his Majesties encouragement, doe sometimes meete together to enquire into and examine (as farr as Philosophie and experience may leade and Conduct them,) all such things as Art, or Nature have produced; that, by a more intimat knowledge and tryall thereof, they may bee able to improve what is allreadie donn, or discovered; or may at least raise by their Inquisition and Industrie some Observations to the benefitt of Mankind, and the advantage of the Commonwealth of Learning.

---

worth noting here that the papers of another member of this committee, Petty, include various documents similar to those discussed here: see the Marquis of Lansdowne (ed.), *The Petty Papers* (2 vols, London, 1927), for example, vol. 1, pp. 175–8, 187–9; vol. 2, pp. 109–10, 115–9. These are hard to date, but most probably date from Petty's later years, like the bulk of his extant manuscripts: see Frances Harris, 'Ireland as a Laboratory: The Archive of Sir William Petty', in Michael Hunter (ed.), *Archives of the Scientific Revolution* (Woodbridge, 1998), pp. 73–91. However, at least one is of much earlier date, 'Enquiries concerning Bathe Waters', evidently dating from *c.* 1650, British Library Add. MS 72892, fol. 1. I am grateful to Adam Fox for drawing my attention to this item.

[57]   Povey to [Digges], 4 March 1660[/61], British Library Egerton MS 2395, fol. 296, with the inquiries referred to in the letter following as fols 297–8, endorsed (fol. 298v), 'The Paper of enquiries from Gresham College to Virginia sent [?] March 4th 1660'. I am extremely grateful to Tina Malcolmson for drawing my attention to this item. A copy of the same letter is to be found in Povey's letterbook, British Library Add. MS 11411, fol. 24: this is endorsed 'Letter to Edward Diggs Esquire Concerning Present of Silke to the King from Virginia' (this refers to the first matter dealt with in the letter), thus identifying the addressee, who is not named in the Egerton MS version (it is dated 'Lincolnes Inne Feilds March the 2d 1660'); however, this copy lacks the accompanying inquiries. Copies of the inquiries survive in Royal Society Classified Papers (hereafter Cl. P.) 19, 65 (in Oldenburg's hand, entitled 'Enquiries of the things peculiar to Virginia and the Bermudas' and with slight differences of wording), and among Abraham Hill's papers in British Library Add. MS 2903, fols 112–13 (titled as in Egerton MS); a printed version derived from them appears in *Phil. Trans.*, 2 (1667), 420–21. The letter in Egerton 2395 is referred to in J.R. Jacob, *Robert Boyle and the English Revolution* (New York, 1977), p. 146, but without referring to the attached inquiries; the Cl. P. and Sloane 2903 copies of the inquiries are referred to in R.P. Stearns, *Science in the British Colonies of America* (Urbana, IL, 1970), p. 694, but without reference to the Povey letter (ibid., pp. 695–8, comprises a different, later set of inquiries for Virginia, again recommended to Edward Digges [1669]).

Povey continued: 'In order to which they have extracted out of such Authors as have writt concerning America a Paper containing some fewe Enquiries, to which a distinct account is desired'. The attached document asked quite specific questions, evidently based on the reading of those who had compiled it, presumably the committee set up in February, who had been instructed to do just this. Concerning earths, for instance, it noted: 'Tis said there is one kinde of a Gummy consistence, white & cleere. another white & so light that it swims uppon water: another red called Wapergh like terra Sigillata', while under fishes it singled out for inquiry 'very large Toadfish. St Georges Dragon. Sting ray with a poisonous preckle'.[58]

This was the first example of a whole genre of inquiries aimed at specific geographical areas, and thereafter similar documents, based on comparably careful scrutiny of appropriate sources, were produced for such places as the East Indies, Iceland, Turkey, Guinea, Egypt, Hungary and Greenland. These were initially enshrined in the society's Register Book and later printed in *Philosophical Transactions* after its inauguration; indeed, in the early months of 1667, two issues of the journal were almost exclusively devoted to such texts.[59] In addition, before his premature death, Lawrence Rooke produced a set of 'Directions for Sea-men, bound for far Voyages', while another important set of inquiries were those concerning agriculture which were prepared by the society's Georgical Committee, one of its most significant institutional initiatives in its early years; these were in fact the first set to be published in *Philosophical Transactions*, and it is worth noting that the primary initiative in producing them seems to have been that of Sir Robert Moray.[60] Later, the production of such inquiries was given

---

58   Egerton MS 2395, fol. 297; Birch, *Royal Society*, vol. 1, p. 17.

59   RBC, vol. 1, pp. 216–21, 287–8; vol. 2, pp. 24–5, 190–91; *Phil. Trans.*, 2 (1667), 344–6, 360–62, 415–22, 467–72, 554–5; 3 (1668), 634–9. See also Cl. P. 19, passim.

60   *Phil. Trans.*, 1 (1666), 91–4, 140–43 (Rooke's 'Directions': see also Birch, *Royal Society*, vol. 1, p. 74, RBC, vol. 1, pp. 153–5). On the agricultural inquiries, see R. Lennard, 'English Agriculture under Charles II: The Evidence of the Royal Society's "Enquiries"', *Economic History Review*, 4 (1932–34), 23–45, and see the discussion in the minutes of the Georgical Committee in Hunter, *Establishing the New Science*, esp. pp. 108–9. See also ibid., pp. 85–6. Prior to Moray producing the inquiries which formed the basis of those circulated, Christopher Merrett had been asked to 'reduce all that belongs to Husbandry to certain comprehensive heads', and William Croone had been asked to extract inquiries from Samuel Hartlib's *Legacie*. The latter reference raises the possibility that the genre of 'heads' owed something to Hartlib, who had published an 'Interrogatory' concerning Ireland in the second edition of his *Legacie* (1652). However, this is arranged in an alphabetical order that none of the Royal Society questionnaires follow. On the 'Interrogatory' and its possible influence, see A.G. MacGregor and A.J. Turner, 'The Ashmolean Museum', in L.S. Sutherland and L.G. Mitchell (eds), *The History of the University of Oxford, Volume 5: The Eighteenth Century* (Oxford, 1986), pp. 636–58, on p. 646n.; Carey, 'Compiling Nature's History', p. 273n., and Charles Webster, *The Great Instauration* (London, 1975), p. 431.

prominence in the account of the society's activities in Thomas Sprat's *History*, which also included a specimen of the responses they elicited.[61]

Of more immediate relevance to Boyle's 'General Heads', there was evidently also a set of general queries for overseas countries. It is clear that such a questionnaire existed at least by January 1662, when 'four or five copies of the inquiries for foreign parts were ordered to be made and given to Mr Povey', while in April 1662, the society's amanuensis:

> was ordered to write out the *Inquiries for foreign parts*, and to deliver them to Mr Boyle; and it was directed, that among these inquiries be inserted one, whether the rain-water varies in weight, trial thereof being made in divers places.[62]

Frustratingly, however, the document in question no longer survives. The only clue as to the possible nature of these inquiries is provided by the general agenda on which the queries sent by Povey to Virginia in March 1661 elaborated, in which case they may have comprised four general categories with more specific queries within them. If so, their content may be reconstructed by taking the 'general' part of each section of the Virginia document, and ignoring queries specific to the region:

1. Concerning the variety of earths.
2. What considerable mineralls, stones, Bitumens, Tinctures, Drugges, & a specimen of each.
   What hot Bathes, & of what medicinall use.
3. What variety of Plants are native there & not in England. what kind of peculiar herbs there are, considerable either for their flower, smell, Alimentary or medicinall use.
4. What kind of animalls, are peculiar to those places.
   1. Insects, flyes, ants, wormes, spiders. Some of each kind to be sent over either alive or dead.
   2. What strange fishes, Tortoises or Turtles.
   3. What Birds
   4. What Beasts.[63]

However, the implication – not least of the instruction concerning the copy for Boyle – is that the document soon became much more elaborate than this, which makes its loss all the more regrettable. The matter arose again in the summer of 1664, when one of the series of committees the society rather

---

[61] Sprat, *History*, pp. 155–72.
[62] Birch, *Royal Society*, vol. 1, pp. 68, 79. Cf. p. 47.
[63] Extracted from Egerton MS 2395, fol. 297.

ambitiously set up at that time, that for 'Correspondence', chaired by Thomas Povey, was ordered to 'consider of drawing up both general and particular heads of inquiries for all the parts of the world'.[64] At its meeting on 19 August 1664, 'Some Generall inquiries, to be sent into all the parts of the world, being red, it was ordered, that the Secretary take care to have them transcribed, and reviewed by some of this Committee, as Sir R. Moray, Dr Wilkins.'[65] However, the Correspondence Committee ceased to meet after 23 September that year, and thereafter nothing further is heard of the matter.

What was Boyle's role in relation to all this? Though, as we have seen, he, with Brouncker, was responsible for the very first document of this kind, the enquiries for Teneriffe produced in January 1661, this was in response to a corporate initiative, and the impression one forms from the society's minutes in these early years is that Boyle's role was a passive rather than an active one in this regard, even concerning the 'general inquiries' whose conception prefigured the famous series with which his name was to be associated. Thus, even if the instruction that he should be given a copy of these in April 1662 was complied with, there is no sequel in the minutes implying that Boyle did anything about it. Similarly, when such matters arose again in the summer of 1664, Boyle was asked for suggestions for inquiries for Guinea, but failed to produce any, while, although he was a member of the Correspondence Committee, he was absent from the meeting when the question of general 'inquiries' was discussed.[66] In a letter of 25 August, Henry Oldenburg informed him:

> we were sorry to be without you, and without your Queries for Guiny. In the mean time, Generall inquiries were drawn up, serving for all the parts of the world; and Authors were distributed amongst the members of this Committee, to be perused for the collecting thence particular inquiries for particular countries.[67]

Boyle's response ignored the matter, while there is also no evidence of his interest in the agricultural inquiries produced at this time by the Georgical Committee, of which Oldenburg sent him a copy in his letter of 1 September, again illustrating that Boyle was peripheral to such initiatives at this point.[68]

Yet within the next eighteen months, Boyle's interest in the genre seems to have been aroused, and it may not be coincidental that this was just the time when *Cold* was being prepared for publication and he evidently came to see the value of 'heads' for the various purposes that have already been itemised. The

---

[64] Birch, *Royal Society*, vol. 1, p. 456. For the committees, see ibid., pp. 397, 402–3, 406–7.

[65] Hunter, *Establishing the New Science*, p. 120. Cf. ibid., pp. 93, 118–20.

[66] Birch, *Royal Society*, vol. 1, p. 454; Hunter, *Establishing the New Science*, pp. 118–20.

[67] *Correspondence*, vol. 2, p. 302.

[68] Ibid., vol. 2, pp. 307–9, 313–14.

context in which he produced his 'General Heads' was the revival of the society's activities in the early months of 1666 after the intermission caused by the plague, when Oldenburg's letters to Boyle repeatedly badger him for material of this kind. In a letter of 27 January, he asked for such guidance that he could publish, reiterating in a letter of 24 February how 'whatever Inquiries you can spare, whether about Insects, or other parts of Natural History, all will be exceeding welcome', and repeating on 6 March: 'what other Inquiries about Naturall things you have ready, and shall think fit to communicate to me for forrain parts, I shall take more than ordinary care to recommend'.[69] Suddenly, in a letter of 21 March, Boyle responded by sending the 'General Heads' to Oldenburg, explaining:

> It belongs to one of the essays of the unpublished part of the Usefulness, &c. and therefore possibly may not so much answer your expectation, as if it had been written entirely for your purpose. But perhaps too, it may serve your turn pretty well, especially with a little addition, which if you make use of it, I can afford it. I have somewhere some specimens of particular enquiries, subordinated to some of the more principal articles of inquisition, which I shall scarce take the pains to look out, till I know, whether the paper I now send may be of use to you. If you have occasion to take any publick notice of it, be pleased to intimate to what treatise it belongs.[70]

The 'General Heads' were published by Oldenburg in issue 11 of *Philosophical Transactions*, dated 2 April 1666.[71] Subsequently, they were followed in issues 18 and 19 in November that year by Boyle's 'Other Inquiries concerning the Sea' and his 'Articles of Inquiries touching *Mines*', which were evidently the 'specimens of particular enquiries' to which he referred in his covering letter to Oldenburg.[72] In publishing these items, Oldenburg capitalised on Boyle's illustrious name, explaining how they had been communicated by 'that lately named *Benefactour to Experimental Philosophy*', and hence drawing on the kudos associated with Boyle of which he and other protagonists of the society made so much. As he put it, alluding to their putative overseas audience in a letter to Boyle of 24 March 1666: 'I know, they will be much pleased and sett on by those comprehensive *Generall Queries*, you give me leave to enrich the next Transactions with.'[73]

---

[69] Ibid., vol. 3, pp. 46, 80, 91. Oldenburg also wished Boyle to add his suggestions for the set of queries sent at this time to Hevelius. See also ibid., vol. 3, pp. 81, 108–9.

[70] Ibid., vol. 3, p. 118. He apologised that 'my haste' made him forget to send it previously as intended. He also responded to the request for additions to the queries to Hevelius, ibid., vol. 3, pp. 117–18.

[71] *Works*, vol. 5, pp. 508–11. They were then read out at a meeting of the Society on 9 May: Birch, *Royal Society*, vol. 2, p. 89.

[72] *Works*, vol. 5, pp. 527–8, 529–40; see above, n. 70.

[73] *Correspondence*, vol. 3, p. 126.

But what can one say about their origins, particularly in the light of Boyle's statement in his covering letter about their belonging to a putative essay for *The Usefulness of Natural Philosophy*? This is not wholly implausible, in that we know that *Usefulness* was going to contain a number of essays that were never published, some of which no longer survive and are known only by lists of contents Boyle drew up. It seems likely that the essay in question was one of these, namely the original sixth essay proposed for the second 'tome' of the book in a synopsis dated 1666, 'That the Naturalist may much advantage Men by exciteing & assisting their Curiosity to discover, take notice, & make use of their homebred Riches & advantages of particular Countrys, & to increase their Number, (by transferring thither those of others)'.[74] Its date is unclear: much of *Usefulness* was written in the 1650s, but some parts dated from the 1660s. If Boyle's composition of this section did precede the Royal Society's initiatives in drawing up such queries, it seems strange that he had not offered this contribution earlier. On the other hand, it is equally likely that this development occurred in parallel with the Royal Society initiative, in connection with another activity in which Boyle was involved at this time, as a member of the Council for Foreign Plantations, to which he was appointed on 1 December 1660. In his autobiographical notes dictated to Gilbert Burnet around 1680, Boyle stated that 'He was of the Committee of forreigne Plantations and in that set himselfe much to know the Natural History of those Countries for which he drew a Paper of Queries'.[75] Since Thomas Povey was very much the guiding force behind the setting up of the Council, it is interesting how closely the queries he sent to Edward Digges in March 1661 echoed the ethos of Boyle's putative essay for *Usefulness*, in their stress on 'those severall kind of things which are reported to be in Virginia & the Bermudas, not found in England'.[76] Possibly Boyle was the mastermind behind both, but it is equally likely that – on this as on other occasions – he claimed the credit for an initiative which was really a shared one.

In fact, it may be argued that these sets of inquiries represent a kind of synergy between Boyle and the Royal Society. In their origins and conception, they unmistakeably belong to the Baconian-inspired programme of information-gathering that was so central to the society's agenda in its earliest years. Yet Boyle's enthusiasm for the genre, once he had discovered it, is patent. Indeed, there are two pieces of evidence which suggest that, once he had developed his enthusiasm for 'heads', Boyle may have been less than fully frank about what had inspired it. One is a comment he made when sending a copy of the sequel to the

---

74 *Works*, vol. 13, p. lxx. See also ibid., pp. lxvi, lxix–lxx; *Boyle Papers*, pp. 114–15, and *Boyle by Himself and His Friends*, pp. lxvii, 29.

75 Ibid., p. 27.

76 See C.M. Andrews, *British Committees, Commissions and Councils of Trade and Plantations, 1622–75* (Baltimore, MD, 1908; reprinted New York, 1970), ch. 4; Egerton MS 2395, fol. 297.

'General Heads', in the form of his 'Articles of Inquiry touching *Mines*', to John Locke on 2 June 1666. In this letter, he alluded to these 'Articles' by a vague formula not dissimilar to that used of the original series in his covering letter to Oldenburg, as a document 'which I once drew up, for the use of some freinds & partly for my owne'.[77] In fact, a heavily revised draft of the published text survives in the hand of an amanuensis who worked for Boyle in the 1660s, and it seems likely that the document in question was of relatively recent composition, notwithstanding the implication of this comment that this was an older text that he had serendipitously come across.[78] It is hard to avoid the suspicion that in this case, as on other occasions, Boyle was deliberately implying that he had long been interested in such things as a means to avoid admitting the extent to which he had been the subject of an external stimulus.

More significant still is an intriguing piece of evidence that Oldenburg did not see the 'General Heads' as so sacrosanct a work by Boyle that it was inappropriate for him to tamper with it – thus implying a sense of ambivalence between the document's being a work by Boyle and the outcome of a more general Royal Society initiative. In 1669, Oldenburg accompanied 'Particular Inquiries for Turky', which he prepared for the English representative at Scanderoon (now Iskenderun), with a manuscript version of the 'General Heads' which is significantly reorganised and clarified compared with the printed version: various unnecessary complexities are omitted, while certain topics were moved from one heading to another where they fitted better.[79] In many respects, its simpler and more logical structure represents an improvement on Boyle's slightly repetitious and digressive original, perhaps indicating the extent to which Oldenburg felt that he knew as well as Boyle what the content of the document should be.[80]

---

[77] *Correspondence*, vol. 3, p. 164.

[78] BP 38, fols 1–5; see *Works*, vol. 5, pp. xxxiv–xxxv, 529–40.

[79] Royal Society Classified Papers 19, 43 (1). See the Appendix to this chapter. In addition, questions concerning the tides were added to sect. 2 from Rooke's 'Directions for Sea-men': see above, n. 60. The attached 'Particular Inquiries for Turky' are endorsed: 'The Inquiries of this paper were recommended to Mr Martyn Lo, Consul (or Vice-Consul) of Scanderoon, who promised to give a fair account to them, Octob. 18. 1669 in London'. For the recipient, Martin Loe, see *Historical Manuscripts Commission, Finch I* (London, 1913), p. 238.

[80] Since this was written, a further relevant piece of evidence has come to light in Robert Hooke's annotated transcript of the minutes of the Royal Society from 1661 to 1677, part of the 'Hooke Folio' sold to the Royal Society by Bonhams on 28 March 2006. Where the minutes record that Boyle's 'General Heads' were read out on 9 May 1666, Hooke has noted: 'Stoln from me'. In fact, no directly comparable text of Hooke's survives, though his 'General Scheme' does list topics for 'particular Histories of the several parts of the World' (Hooke,

In general, however, it was Boyle's own version which was given wide circulation through its publication in print in *Philosophical Transactions* in 1666, and it achieved an extraordinary longevity. First, it was reprinted in 1692 in a volume comprising this and other questionnaires from the early volumes of *Philosophical Transactions* which was given an overall title echoing Boyle's: *General Heads for the Natural History of a Country, Great or Small*. Indeed, in this instance, the publisher attributed the whole book to Boyle on its title page, notwithstanding the fact that it really reflected the corporate activity of the Royal Society in this respect, including material by such authors as Lawrence Rooke, Thomas Henshaw, Sir John Hoskins, Charles Howard and Oldenburg.[81] Thereafter, versions of Boyle's 'General Heads' continued to appear in an almost emblematic position in collections of travel writings into the eighteenth century, for instance in Jean-François Bernard's *Receuil des voyages au nord* of 1715–27.[82] This has often and rightly been seen as a tribute to Boyle's influence on the development of natural history in the eighteenth century, but it is proper to acknowledge the extent to which Boyle might never have gone down this route but for the influence of the Royal Society.[83]

Once launched, the genre of 'inquiries' popularised as a result of these activities by Boyle and the Royal Society in the 1660s became more widespread. From the 1670s onwards, we encounter a whole series of specially printed 'inquiries' issued as separate broadsheets which, as various scholars have noted, played a significant role in the natural philosophical (and antiquarian) enterprise of the period.[84] Thus the publisher John Ogilby issued *Queries in Order to the Description of Britannia* in 1673, and Robert Plot, first Curator of the Ashmolean

---

Posthumous Works, pp. 22–3). Here we evidently see Hooke's notorious possessiveness at work in relation to ideas which were in fact widely shared in the early Royal Society.

[81]   See *Works*, vol. 5, pp. xli–xlv. In this connection, it is interesting that it was to Boyle that John Houghton attributed the Society's agricultural inquiries when he reprinted them in *A Collection of Letters for the Improvement of Husbandry and Trade* (2 vols, London, 1681–3), vol. 2, pp. 81–2 (alluding to the reprint of the inquiries in ibid., vol. 1, pp. 6–9).

[82]   J.-F. Bernard, ed., *Receuil des voyages au Nord* (8 vols, Amsterdam, 1715–27), vol. 1, pp. 2–5; vol. 4, p. lxii.

[83]   See, for example, P.J. Marshall and Glyndwr Williams, *The Great Map of Mankind* (London, 1982), p. 45; Anita Guerrini, *Natural History and the New World, 1524–1770* (Philadelphia, PA, 1986), p. 5; Charles Withers, 'Geography, Science and National Identity in Early Modern Britain: The Case of Scotland and the Work of Sir Robert Sibbald', *Annals of Science*, 53 (1996), 29–73, on p. 63; Richard Drayton, *Nature's Government: Science, Imperial Britain and the 'Improvement' of the World* (New Haven, CT and London, 2000), p. 17.

[84]   See, for instance, Withers, 'Geography, Science and National Identity'; Michael Hunter, *John Aubrey and the Realm of Learning* (London, 1975), pp. 71–2; S.A.E. Mendyk, *Speculum Britanniae* (Toronto, 1989), ch. 11; Daniel Woolf, *The Social Circulation of the Past* (Oxford, 2003), p. 159.

and author of *The Natural History of Oxfordshire* and of *Staffordshire*, brought out two such sets of printed inquiries in 1674 and 1679.[85] Plot was followed by his successor at the Ashmolean, Edward Lhuyd, whose *Parochial Queries* for Wales appeared in 1696, by which time the practice had spread to Ireland in the hands of William Molyneux and to Scotland in the hands first of Sir Robert Sibbald and then of Robert Wodrow.[86] Meanwhile, in England, we find such publications of a similar type as John Woodward's *Brief Instructions for Making Observations in All Parts of the World*, issued under the imprimatur of the Royal Society in 1696.[87] In some cases, we have sets of the responses which such queries elicited, and as a whole the genre has only begun to receive the full study which it richly deserves.[88]

In conclusion, however, it is appropriate to revert to Boyle. What I hope I have established here is that the Baconian genre of 'heads', which was subsequently to prove so central to Boyle's natural philosophical method, came into his corpus at a specific point, in the 1660s, and that by far the likeliest stimulus to this was the influence of the Royal Society, in whose early corporate activity the compilation of comparable documents played such a central role. This is symptomatic of a more general need to look beyond Boyle's own *oeuvre* in order properly to understand him. Another instance of this, also involving the Royal Society, is provided by Boyle's novel practice in the early 1670s of publishing his writings in the form of brief 'Tracts', which was almost certainly inspired by his experience of publishing in *Philosophical Transactions* in the 1660s.[89] Again, this is something one can only divine by observing a shift in Boyle's behaviour and seeking a plausible explanation for it in his milieu. In the case of the 'heads' and 'inquiries' on which I have focused here, one sees a revealing reciprocity, which is all the more significant because of the influence enjoyed by the genre for which

---

[85] A copy of the earlier set of Plot's enquiries can be found in Cl. P. 19, 93, and of the revised set in ibid., 94; the latter is appended to Paul Minet's reprint of Plot's *Natural History of Oxfordshire* (Chicheley, 1972).

[86] K.T. Hoppen, *The Common Scientist in the Seventeenth Century* (London, 1970), pp. 21–3, 200–201; Withers, 'Geography, Science and National Identity', esp. pp. 66–73; Michael Hunter, *The Occult Laboratory: Magic, Science and Second Sight in late Seventeenth-century Scotland* (Woodbridge, 2001), esp. pp. 23ff.

[87] Reprinted., with intro. by V.A. Eyles, London, 1973.

[88] See esp. R.H. Morris (ed.), *Parochialia, being a Summary of Answers to 'Parochial Queries... issued by Edward Lhuyd*, three supplementary volumes to *Archaeologia Cambrensis*, 1909–11. For the responses to Molyneux's inquiries, see Hoppen, *Common Scientist*, p. 234 n. 101; for Scotland, see the references cited above, n. 86. For a valuable study of the genre, see Adam Fox, 'Printed Questionnaires, Research Networks, and the Discovery of the British Isles, 1650–1800', *Historical Journal*, 53 (2010), 593–621. I am grateful to Dr Fox for allowing me to read a draft of this article prior to publication.

[89] See *Works*, vol. 1, pp. xxxvi–xxxviii.

Boyle and the Royal Society were jointly responsible. But there is also a broader lesson concerning the relationship between biography and institutional history. No man is an island: even Boyle, for all his self-preoccupation, took lessons from his peers. The challenge is to integrate the subject with his setting, and the result is to enrich our understanding of both.

Appendix
Text of Oldenburg's Version of Boyle's 'General Heads'
(Royal Society Classified Papers 19, 43 (1))[1]

General Heads of Inquiries for all Countries.

1. Concerning the *Air*;

What is the usual salubrity and insalubrity of the Air?

What diseases the Country is most subject to?

What are the Variations of the Weather, according to the Seasons of the year, and the times of the day? And what Duration the severerall [*sic*] kinds of Weather usually have?

What Meteors it is most wont to breed; especially, what Winds it is subject to; whether any of them be stated and ordinary etc.?

2. About the *Water*;

What is the Depth of the *Sea*, its degree of Saltnes,[2] Currents, Tydes? And, as to the Tydes, what is their precise Time of Ebbing and Flowing in Rivers, at Promontories or Capes; which way their Current runs; what Perpendicular distance there is between the highest Tyde and lowest Ebbe, during the Spring-Tides and Neap-Tides? What are the degrees of the Risings and Fallings of the Water in *Equal* spaces of time and the *Velocity* of its motion at several heights? What day of the Moons age, and what times of the year, the highest and lowest Tides fall out etc.?

For *Rivers*; What is their bignes, Length, Course, Inundations, Goodnes, Levity of waters?

For *Lakes, Ponds, Springs*, and especially Mineral waters, their kinds, Qualities, Vertues, and how examined?

For all sorts of Waters; what kinds of *Fishes* they breed; their store, bignes, goodnes, seasons, haunts, peculiarities of any kind, and the ways of taking them?

---

1   This is fol. 83 of the volume; the conjugate leaf, fol. 84, comprises 'Particular Inquiries for Turky': see above, p. 75 n. 79.

2   Followed by 'Tydes' deleted. Six words later, 'the' was inserted in the space between the adjacent words after composition.

3. About the *Earth*;

　[3]Whether ‹the Country be› plain, or mountanous, or both? If montanous, what is the height of the tallest mountains? Whether they lye scatter'd, or in ridges; and whether those run North and South, or East and West etc.? What Promontories, /verso/ fiery or smoaking Hills the Contry has or hath not?

　What the Magneticall Declination is in several places, and the Variation of that declination in the same place?

　What the nature of the Soyle is, whether Clayie,[4] Sandy, etc. or good Mould? And what Grains, fruits and other vegetables doe the most naturally agree with it; and especially, what Fruit- and Timber-trees,[5] and what other Trees, whose wood is considerable, the Contry has or wants? By what particular Arts and Industries[6] the Inhabitants improve the advantages, and remedy the Inconveniences of their Soyle?

　What *Animals* the Contry has or wants, both wild and tame? And as to the Inhabitants, Men and Woemen, What is their Stature, Shape, Colour, Features, Strength, Agility, Beauty, Dyet, Inclinations? What the Fruitfulnes or Barrenes, hard or easy Labour of the Woemen etc.?

　What Minerals the Contry is stored with? What Quarries it affords; and the particular conditions both of the Quarries and the Stones; as also, how the Beds of Stone lye, in reference to North and South etc.? What Clays and Earths it affords, as Tobacco-pipe Clay, Marles, Fullers-earths, Earths for Potters-wares, Bolus's and other Medicated Earths? What other[7] Minerals it yields, as Coals, Salt-mines, or Salt-springs, Allom, Vitriol, Sulphur etc.? What Mettals[8] it affords; and a description of the Mines, their number, situation, depth, signs, waters, damps, quantities of ore, goodnes of ore, the ways of reducing their oares into mettals etc.?

Adde hereunto the Names of the Men, excelling at the present in Philosophicall, Mathematicall, Mechanicall, Medicall and Chymicall knowledge, together with the Books, they have published or shall[9] publish from time to time.

---

　　[3]　Preceded by 'As to its Figure' deleted.

　　[4]　Altered from 'Clays'. The next word, 'Sandy', is altered from 'Sands'.

　　[5]　Followed by '(esp' deleted.

　　[6]　Followed by 'impro' deleted.

　　[7]　Followed by 'What' deleted, evidently when 'What other' inserted in space at the end of the previous line.

　　[8]　Followed by 'the' deleted.

　　[9]　Altered from 'are' and followed by 'now' deleted. The next word, 'publish', is altered from 'publishing'. This paragraph was added to the text after composition in darker ink.

Chapter 4

Boyle, Narcissus Marsh and the Anglo-Irish Intellectual Scene in the Late Seventeenth Century[1]

In this chapter, I want to use the career of Boyle, as an Irish-born natural philosopher, as a kind of prism to examine various aspects of Anglo-Irish relations during the second half of the seventeenth century. I will focus on the shared concerns that brought Boyle together with Narcissus Marsh in the 1680s, drawing on the 2001 edition of Boyle's *Correspondence*, in which many of the letters between these two men were published for the first time.[2] In particular, I will deal with the project for publishing the Bible in Irish that preoccupied Boyle and Marsh in these years, but I will also say a little about their other mutual interests – thus, I hope, shedding a small amount of new light on the background to the founding of the Dublin Philosophical Society in 1683.

Boyle was the fourteenth child and seventh son of Richard Boyle, the 'Great' Earl of Cork. Born at Lismore Castle on 25 January 1627, he was brought up in Ireland for the first eight years of his life, until he was sent to be educated at Eton College in 1635. In fact, his earliest recollection, recounted by the biographer John Aubrey in his *Brief Life* of Boyle, is of a very Irish experience: 'He was nursed by an Irish nurse, after the Irish manner, wher they putt the child into a pendulous satchell (insted of a cradle), with a slitt for the child's head to peepe out.'[3] Evidently recalling such memories, Boyle was still talking of 'us in Ireland' in 1647 in connection with:

> the Fondnesse & Concerne generally observ'd in Nurses, towards those they have
> been related to in that quality; & on the Foster-Childe's part, by the Kindnesse
> usuall amongst us in Ireland (where Good Nature is permitted to act freely,

---

[1] Previously published as 'Robert Boyle, Narcissus Marsh and the Anglo-Irish Intellectual Scene in the Late Seventeenth Century', in Muriel McCarthy and Ann Simmons (eds), *The Making of Marsh's Library: Learning, Politics and Religion in Ireland 1650–1750* (Dublin: Four Courts Press, 2004), pp. 51–75.

[2] *Correspondence*, esp. vols 5 and 6.

[3] Aubrey, *Brief Lives*, ed. Andrew Clark (2 vols, Oxford, 1898), vol. 1, p. 120.

without Danger of passing for any thing besides it selfe,) both to those that have furnish't us with their Milke, & to those that have suck't it with us.[4]

As his life progressed, however, Boyle's experiences increasingly distanced him from his country of birth, to which he returned only briefly after leaving it as a child. He recorded that 'I could never yet be induced to learn the native tongue of the kingdom I was born and for some years bred in', and he might have come to exemplify the attitude which he expressed in a letter to his friend John Mallet in 1653 of Ireland as 'a Country, which those that have the most Relation to it, seldome thinke on any longer then they are in it'.[5]

Yet Boyle's entire life was inextricably tied to Ireland, not least due to the fact that his lifestyle was made possible by his ample endowment with lands there by the Earl of Cork, whose bequests to his sons were on as lavish a scale as his tomb at St Patrick's Cathedral (on which Boyle appears as a child, the earliest depiction of him).[6] Boyle's patrimony comprised extensive estates scattered throughout Ireland, together with the income from impropriated livings, formerly owned by monastic foundations dissolved at the Reformation.[7] These provided a large income for Boyle throughout his life – said at the time to be £3,000 per annum – though there were problems in the aftermath of the Civil War and again in 1688–89, when the disturbances in Ireland threatened

---

[4]   Robert Boyle, 'The Duty of a Mother's being a Nurse, asserted' (dated 15 August 1647), in *Works*, vol. 13, pp. 71–2. For a putative trip to Ireland in 1649 which failed to materialize, see *Correspondence*, vol. 1, pp. 85–7.

[5]   Section of Boyle's 'Essay of the Holy Scriptures' (*c.* 1652–54), in *Works*, vol. 12, p. 357; *Correspondence*, vol. 1, pp. 139–40. Note also Boyle's description of himself as 'so great a stranger to the Irish Language' in *Correspondence*, vol. 5, p. 378, and see T.C. Barnard, 'Protestants and the Irish Language, c. 1675–1725', *Journal of Ecclesiastical History*, 44 (1993), 243–72, on p. 247, which cites the reference just given, and R.E.W. Maddison, *The Life of the Hon. Robert Boyle* (London, 1969), p. 11n., where a letter is quoted showing that Boyle and his brother's tutor at Eton tried to interest them in Irish, and Boyle 'Some tymes desirs it and is a litle intred in it'; but his overall verdict was that 'they affect not the Irish, notwithstanding I shew many reason to bind their minds thereto'.

[6]   See Amy L. Harris, 'The Funeral Monuments of Richard Boyle, Earl of Cork', *Church Monuments*, 13 (1998), 70–86, on pp. 82–3 and fig. 21 (though Boyle's Christian name is wrongly given as Roger).

[7]   See Dorothea Townsend, *The Life and Letters of the Great Earl of Cork* (London, 1904), pp. 468ff., on pp. 485–8. See also National Library of Ireland MS 6244. The complementary information available from rentals from Cork's later years in the National Library of Ireland is summarised in *Between God and Science*, p. 310 n. 51. See also Chatworth House, Lismore MS 33, items 96–142, the bulk of which comprise papers relating to Boyle's estates, mainly in the early 1680s, including receipts for 1682–83 (117–18). The sense that these give of the scale of his income bears out the estimate given by Aubrey (see below, n. 8).

disaster.[8] Indeed, at the latter point – and still in his will, drawn up in 1691 – Boyle was worried whether he would be able to fulfil his commitments due to his loss by 'the destructive Insurrections and Warr that hath hapned in Ireland the whole Incomb for above Two yeares Last past of my Estate there'.[9]

All this provides a crucial subtext to Boyle's life, in that throughout his career there is a kind of 'white noise' of negotiations with agents and with his siblings and others concerning the detailed administration of his lands and the income derived from them. In addition, and partly in this connection, Boyle formed part of a kind of Anglo-Irish establishment, involving not only his brothers and sisters and their extended families, but also other landed figures such as Sir Robert Southwell and Sir John Percivall. Indeed, doubtless through such contacts, Boyle was able to dilate in a well-informed way on Irish affairs from his London base long after his last visit to the country, as is illustrated by the comment of Sir Paul Rycaut in a letter to Boyle following his arrival in Ireland in 1686 to take up office as secretary to the Lord Lieutenant: 'the account you gave me of the state of this country one day at a private conference together, I have often remembered, and compared it with the disposition of this country, and find it very agreeable, and corresponding in all particulars'.[10]

On the other hand, Boyle's association with an Anglo-Irish milieu in London has led to some misunderstanding, not least concerning his links with the circle surrounding the intelligencer Samuel Hartlib, from whom Boyle received profuse letters throughout the 1640s and 1650s. In particular, as we saw in Chapter 2, there has been misapprehension about the so-called 'Invisible College' to which he belonged in the late 1640s, the evidence for which comprises a few brief hints in extant letters. On the basis of these, Charles Webster some years ago visualised the 'college' as a kind of Anglo-Irish confraternity, comprising Boyle, his sister, Lady Ranelagh, and such associates of theirs as Benjamin Worsley; Webster saw these as promoting a programme of inquiry, experiment and improvement, foreshadowing the concerns of the mature Boyle and having specific reference to Ireland.[11]

Unfortunately, as has been shown above, this picture is almost wholly fanciful. There is no evidence that the Invisible College was concerned with experiment and improvement at all: if anything, it seems likelier to have been concerned with the reflections on moral issues to which Boyle largely devoted himself in the late 1640s, before the almost Pauline conversion experience by which he discovered science in the years following 1649. There is even less

---

[8]   Aubrey, *Brief Lives*, vol. 1, pp. 120–21. For the problems in 1646–7, see *Correspondence*, vol. 1, pp. 29–30, 49–50.

[9]   Maddison, *Life of Boyle*, p. 267. Cf. *Correspondence*, vol. 6, p. 297.

[10]   Ibid., vol. 6, p. 167. For a much earlier appraisal, see Boyle's letter of 22 October 1646 to Isaac Marcombes in ibid., vol. 1, pp. 37–8.

[11]   For full documentation, see above, pp. 46–8.

evidence that it had anything at all to do with Ireland. Insofar as there *was* a formative Anglo-Irish influence on Boyle in these years, it was rather different, in the person of James Ussher, Archbishop of Armagh and a family friend, who was responsible for a parallel development in Boyle's interests at the time when he first discovered science in the form of his discovery of the importance of philology and erudition, aimed particularly at the accurate interpretation of the Bible. Of Ussher, Boyle wrote: 'whose encouragements I gratefully acknowledge to have much engaged me to the study of the holy tongues', explaining how the eminent divine 'reproaching him that was so studious for his ignorance of the Greek he studied it and read the New Testament in that Language so much that he could have quoted it as readily in Greek as in English'. From this, as was explained in Chapter 2, Boyle went on to the study of Hebrew, Chaldaic, Syriac and Arabic, even writing a grammar of Hebrew for his own use.[12]

The significance of this for us is not least the fact that when Boyle returned to Ireland for the first (and last) time after his childhood, in 1652–54, his main preoccupation seems to have been biblical study. He explained later that it was at this time he wrote the 'Essay of the Holy Scriptures' that was to be partially published as *Some Considerations touching the Style of the Holy Scriptures* in 1661. Moreover, that his principal learned interest on the visit was theological is suggested by the fact that the only books he requested from Hartlib were 'Arminian' ones, the exact nature of which is unfortunately unclear.[13]

The other topic with which we know that he concerned himself on this trip was anatomy, which he explored with the adventurer and polymath William Petty, whom he had met in 1648 through Hartlib, and who was now in the first phase of his effervescent career in Ireland. A telling letter survives from Petty to Boyle, written during Boyle's Irish stay, discussing medical matters and criticising his valetudinarianism, while we also learn from a letter from Boyle to Hartlib's son-in-law, Frederick Clodius, in the spring of 1654, that he was:

> exercising myself in making anatomical dissections of living animals: wherein (being assisted by your father-in-law's ingenious friend Dr *Petty*, our general's physician) I have satisfied myself of the circulation of the blood, and the (freshly

---

[12] *Works*, vol. 12, pp. 355–6; *Boyle by Himself and His Friends*, p. 27. All this was, of course, after Ussher had left Ireland in the aftermath of the Civil War: Boyle initially records contact with Ussher in 1646 (*Correspondence*, vol. 1, p. 40), by which time Ussher was living in exile in Wales, and Boyle may have met him there or on a periodic foray to London. On Boyle's links with Ussher, see also Chapter 2 in this volume and *Scrupulosity and Science*, pp. 33, 43–4.

[13] *Works*, vol. 2, pp. xxv–xxvi, 387; vol. 13, pp. xxxix–xl. *Correspondence*, vol. 1, pp. 158, 169. It is perhaps also worth noting that Boyle's letter to Mallet of January 1653 is almost entirely about biblical matters, though this could be because that is what Mallet had asked him about: ibid., vol. 1, pp. 139–41. See also ibid., vol. 1, p. 134.

discovered and hardly discoverable) receptaculum chyli, made by the confluence of the venæ lacteæ; and have seen (especially in the dissections of fishes) more of the variety and contrivances of nature, and the majesty and wisdom of her author, than all the books I ever read in my life could give me convincing notions of.[14]

This provides interesting evidence that the very latest ideas in physiology were circulating in the circles in which Boyle moved in Dublin, which included not only Petty but also Dr Robert Child and others. On the other hand, Boyle's reaction to Ireland during this trip was rather negative. He described it as a 'barbarous', 'illiterate' country, where 'I am kept prisoner', complaining of the lack of facilities for chemical experiments and the lack of appropriate company. Indeed, he almost certainly over-complained, as Hartlib tactfully pointed out in a letter to Boyle of 28 February 1654.[15]

It is perhaps equally interesting – in view of the unity of purpose that is sometimes imputed to the Hartlib circle, and the significance that is attributed to its schemes – that Boyle seems to have failed entirely to rise to Hartlib's plea to him to help in the scheme for a Natural History of Ireland that had been begun by Hartlib's associate Gerard Boate prior to his early death in 1650. Hartlib sent Boyle multiple copies of the *Interrogatory Relating more particularly to the Husbandry and Naturall History of Ireland* that he had printed as an annexe to the second edition of his *Legacie: or An Enlargement of the Discourse of Husbandrie Used in Brabant & Flaunders* in 1652. He fulsomely described the task and its import, sending various queries 'worth a philosophical pen in these places', the resolution of which would mean 'that so by little and little we might perfectly come to understand the natural history of all the parts in that country'. 'I suppose', he wrote, 'this may be one means, whereby *Ireland* may be peopled again, and get good tenants; especially, if the other parts, which are wanting to that history, were more particularly discovered and described.' Indeed, in his letter to Boyle of 28 February 1654, he averred: 'I must now most solemnly call upon you, on the behalf of the *Natural History of Ireland*, which, if yourself and Dr *Child* do not take professedly to task, I fear will never be perfected to any purpose.'[16]

Boyle, however, seems not to have risen to the challenge at all. Even such information as he relayed to Hartlib was cursory, due to 'my perpetuall distractions', and, though he made a few observations which he was to record in his own works over the following decade, he completely failed to assist Hartlib

---

[14] Ibid., vol. 1, pp. 142–4, 167. See also ibid., vol. 1, p. 177.

[15] Ibid., vol. 1, pp. 166–7. See also Hartlib's paraphrase in ibid, vol. 1, p. 159. For a later, and slightly disingenuous, invocation of the backward state of Ireland on Boyle's part, see *Works*, vol. 3, p. 18.

[16] *Correspondence*, vol. 1, pp. 157–8, 169–70, 176–7.

in his project.[17] Hartlib may have been unlucky in his Irish contacts, as seen most
notoriously in the infighting between Petty and Worsley,[18] but it is revealing
that Boyle, too, disappointed him. In fact, I am sorry to have to say that Boyle's
Irish experience at this point was quite negative. In writing to Clodius, he made
clear his hope 'that by a short and necessary stay, to settle my affairs in this
country, I may put myself in a condition of living out of it, and prosecute more
undistractedly and effectually the study of real learning'.[19] To make things worse,
he fell seriously ill towards the end of his stay. As he himself wrote much later, in
the preface to the second volume of his *Medicinal Experiments*, posthumously
published in 1693:

> The grand Original of the Mischiefs that have for many Years afflicted me, was a
> fall from an unruly Horse into a deep place, by which I was so bruised, that I feel
> the bad Effects of it to this day. For this Mischance happening in *Ireland*, and I
> being forc'd to take a long Journey, before I was well recovered, the bad Weather
> I met with, and the as bad Accommodation in *Irish* Inns, and the mistake of an
> unskilful or drunken Guide, who made me wander almost all Night upon some
> Wild Mountains, put me into a Fever and a Dropsie, (*viz.* an *Anasarca:*) For a
> compleat Cure of which I past into *England*, and came to *London*.[20]

Perhaps it is not surprising that he never stepped on Irish soil again.

In the late 1650s and early 1660s, Boyle was at Oxford, part of the famous
Experimental Philosophy Club which met there in those years, and it was at this
time that he first performed and then wrote up and published the experiments
that were to make him famous. In such books as his *Certain Physiological
Essays* (1661) and *The Usefulness of Natural Philosophy* (1663), he made passing
references to his Irish experiences, showing that he had kept his eyes open while
in Ireland despite failing to oblige Hartlib with an account of its natural history.[21]
In addition, in these years, as at other times in his life, Boyle continued to be
reminded of Ireland by a steady flow of business that had to be executed, which

---

[17]   Ibid., vol. 1, pp. 168–9. For Boyle's observations, see below, n. 21.

[18]   See Charles Webster, *The Great Instauration: Science, Medicine and Reform 1626–60*
(London, 1975), pp. 75–6, 434–7; T.C. Barnard, *Cromwellian Ireland* (Oxford, 1975),
pp. 229–33.

[19]   *Correspondence*, vol. 1, p. 166.

[20]   *Works*, vol. 12, p. 211. In fact, Boyle's problems did not end on his arrival in London,
as he goes on to explain.

[21]   See ibid., vol. 2, pp. 74, 80; vol. 3, pp. 306–7, 339, 444, 464. For references in writings
published at a later date, see ibid., vol. 6, pp. 315, 429; vol. 7, pp. 250, 296; vol. 10, pp. 309,
321, 334, 366; vol. 13, pp. 206–7, 231, 393.

occasionally rose to a climax, as with the parliamentary debates over legislation to prevent the import of Irish cattle into England in the mid-1660s.[22]

Ireland also occasionally arose in the proceedings of the newly founded Royal Society in the 1660s, not least in connection with the society's short-lived preoccupation in its earliest phase with matters of immediate economic significance. One such project was William Petty's 'double-bottom', a catamaran-style boat described by the society's historian, Thomas Sprat, as 'the most considerable *Experiment*, that has been made in this *Age* of *Experiments*': the results of the trials of the prototypes of this in Dublin Bay were reported at various meetings of the society in 1662–63.[23] Similarly, when the society took an interest in the cultivation of potatoes as a means of alleviating the suffering in the English countryside caused by a succession of bad harvests:

> Mr Boyle related, that he knew, that in a time of famine in Ireland there were kept from starving, thousands off poor people, by potatoes: And that this root would make good bread, mixed with wheaten-meale: that it will yield good drink too, but of no long duration: that it will prosper after the plow, as well as by delving: that it feeds poultry, and other animals well: that any refuse will keep them from frost: that the very stalkes of them, thrown into the ground, will produce good roots: that the planting of them doth not hinder poor people from other imployment.[24]

There is also the notorious affair of Valentine Greatrakes, the Irish 'stroker' who came to England and effected marvellous cures there in the early months of 1666. Boyle became quite closely involved with Greatrakes, compiling a set of queries concerning his healing powers, carefully observing a series of cures by him, and even experimenting with the use of Greatrakes' glove to see if he could achieve similar therapeutic effects himself.[25]

---

[22]   See esp. *Correspondence*, vol. 2, 560ff.

[23]   Thomas Sprat, *The History of the Royal Society* (London, 1667), p. 240; Birch, *Royal Society*, vol. 1, 124, 131, 141, 180, 183–92, 194, 249, 279, 287, 310; the Marquess of Lansdowne (ed.), *The Double Bottom or Twin-hulled Ship of Sir William Petty* (Oxford, 1931), pp. 25ff.

[24]   Quoted in Michael Hunter, *Establishing the New Science: the Experience of the Royal Society* (Woodbridge, 1989), pp. 103–4. See also ibid., pp. 77–8.

[25]   *Correspondence*, vol. 3, pp. 82ff., passim. For Boyle's queries, see C.S. Breathnach, 'Robert Boyle's Approach to the Ministrations of Valentine Greatrakes', *History of Psychiatry*, 10 (1999), 87–109, and Michael Hunter (ed.), *Robert Boyle's 'Heads' and 'Inquiries'*, Robert Boyle Project Occasional Papers No. 1 (London, 2005), pp. 31–2. For Boyle's notes on Greatrakes' cures (not known to Breathnach), see British Library Add. MS 4293, fols 50–53, now published as no. 26 of Boyle's workdiaries, http://www.livesandletters.ac.uk/wd/index.html. See further below, pp. 169–70.

More significant was the Restoration settlement in Ireland, which must have had the effect of reinforcing the slightly mixed image of Ireland that Boyle already had. As part of the Restoration settlement, Charles II granted Boyle a 31-year lease of the impropriations of various former abbey lands, and this episode had implications which are worth examining. Now the land settlement, following on from the redistribution that had occurred under the aegis of Petty in the 1650s, was clearly seen by many as a great opportunity. Thus Boyle's cousin, Michael Boyle, Bishop of Cork, sought to have money put aside from it for the better endowment of the Irish church. Equally significant, in view of the earlier Hartlibian interest in Ireland, was the fact that in 1662–63 the Royal Society made a bid for funding from this source: an application was made by Boyle and the courtier Sir Robert Moray for the society to be granted fractions accruing from the land settlement; in addition, the planters and adventurers in Ireland were invited to contribute to the society's design.[26] In fact, like all the society's attempts to gain a revenue in its early years, this proved abortive: instead, the society's sole source of income was the subscriptions of its Fellows, including, of course, the Anglo-Irish contingent to which Boyle belonged.[27]

If the Royal Society failed to benefit from the Restoration settlement, however, Boyle did not. An application was made on his behalf – whether with or without his consent is unclear – for a grant of impropriations, and this was indeed successful, though it proved more difficult to implement than had been anticipated, and it was over a decade before Boyle enjoyed the proceeds.[28] Even at the time, complaints arose over its legitimacy, since Boyle was challenged by his cousin, Bishop Boyle, who told him that, though satisfied that Boyle's motives were 'charitable and relligious':

> yet I must humbly take leave to acknowledge my selfe unsatisyed why that additionall revenue which his Majesty designed & promised for the better

---

[26] See *Calendar of State Papers relating to Ireland, 1660–2*, pp. 602, 668; *1669–70*, pp. 428–9; BP 40, fol. 4; Royal Society Domestic MSS 5, nos 34–6; Birch, *Royal Society*, vol. 1, pp. 168–9; A.R. and M.B. Hall (eds), *The Correspondence of Henry Oldenburg* (13 vols, Madison, WI and London, 1965–86), vol. 2, pp. 48–50, 52–5.

[27] See Michael Hunter, *The Royal Society and its Fellows 1660–1700: The Morphology of an Early Scientific Institution* (new edn, Oxford, 1994), passim.

[28] See *Boyle by Himself and His Friends*, pp. lxxiii, 27–8, 78; *Scrupulosity and Science*, p. 75; Edward MacLysaght, *Calendar of the Orrery Papers* (Dublin, 1941), p. 18; British Library MS Egerton 2549, fol. 96; Maddison, *Life of Boyle*, pp. 98, 110, 271n. On the protracted nature of the outcome, see *Calendar of State Papers, Ireland, 1666–9*, p. 255 (confirmation dated 19 December 1666); *Calendar of State Papers Domestic, 1671*, pp. 278–9; *Correspondence*, vol. 3, pp. 289–90, and vol. 4, pp. 198ff., passim; and Bodleian Library, Oxford, MS Rawlinson B 492, a document in a volume of royal letters dated 10 January 1673 giving remission of rent due to the fact that Boyle did not gain possession till about May 1669.

support of the Clergy here that they may with greater comforte attend the cures of their severall churches should be diverted to any other use though in it selfe it be generous & hansome. Especially when I consider how many congregations depend uppon the service of those parishes; how destitute they will be of a Church to resorte unto, how the poore people will be compeld to wander through the Countrey to find out an opportune place for the performance of there publick dutyes & devotions unto God.[29]

Perhaps partly for this reason, Boyle's strategy for the use of the revenue from this source changed. Initially, he seems to have intended to use the money 'for the Charge he was at in his laboratory' – in other words, to subsidize the intensive experimental work that he had by this time put in hand and which saw the light of day in the treatises already referred to, exemplifying Boyle's retrospective importance in the history of science. But although, in his letter to Bishop Boyle, 'the Advancement of reall Learning' was one of the uses to which he vaguely claimed he intended to put the money, in justifying why he, rather than the Irish church, was the proper beneficiary of this largesse, Boyle laid much more stress on a different use. This was linked to a new responsibility he had taken up at the Restoration, as Governor of the Corporation for the Propagation of the Gospel in New England, or New England Company. This was the principal example of the involvement in and sponsorship of missionary activity that Boyle was to engage in for the rest of his life. Citing the needs of the New England Company in justifying the grant to him, he argued that 'soe pious a Designe as is pursu'd by this Corporation is now in danger to miscarry for want of Maintainance'.[30]

Yet I think Boyle was genuinely bruised by this altercation with his cousin, and it is worth pointing out here that he was in any case susceptible to anxieties concerning such matters since he was of a highly 'scrupulous' disposition, in the literal sense of that word, spending hours worrying about the legitimacy of his actions, and seeking the advice of churchmen to help him ease his conscience on matters of this kind. We know that he consulted one of his casuistical advisors, Bishop Robert Sanderson, on the issue of these impropriations, described by Boyle as 'soe nice an occasion', while among the most interesting extant documents associated with Boyle are some actual notes on interviews on related matters with prominent bishops in the last year of his life, which I have published elsewhere.[31] Indeed, it could be argued that, so far from the triumphalist, rather manipulative view that has characterized some accounts of

---

[29] *Correspondence*, vol. 2, pp. 41–2.

[30] Ibid., vol. 2, pp. 23–4: Boyle also stated that he would support the ministers and the poor in the areas covered by the impropriations. For the explicit link with 'the Charge he was at in his laboratory', see *Boyle by Himself and His Friends*, p. 28.

[31] *Correspondence*, vol. 2, p. 23; *Scrupulosity and Science*, pp. 87–92.

Boyle and his links with Ireland and the wider world, in fact what we see is an extraordinarily contorted attitude, not least in his relations with the country of his birth and the income he derived from it.[32]

Although such issues were raised acutely by the impropriations granted under the Restoration settlement, similar problems arose with the impropriations Boyle had inherited from his father. Hence the next well-documented episode when Boyle was in frequent contact with Ireland concerned the ministers serving the livings covered by these, for whom he sought to make financial provision with the help of his contact Robert Southwell of Kinsale. This took place in the early 1670s, and it may not be coincidental that this occurred just as the complications over the impropriations granted in the Restoration settlement were finally being sorted out and Boyle began to enjoy a revenue from them: arguably, he could now more easily afford to salve his conscience on this matter than hitherto. Boyle used Southwell to make enquiries on his behalf about the deservingness or otherwise of the different clergy involved, and he then arranged for differential support to be given them on the basis of the information he received, marking a schedule with 'a black Lead pen' as to exactly which minister was to get what.[33] However, this was probably only part of Boyle's largesse towards the ministers in his impropriated parishes, not all of which is documented at all, since much of his charitable giving was conciously kept secret. Thomas Birch, Boyle's earliest biographer, recorded that 'a person who was concerned in two distributions which were made, declared, that the sums upon those two occasions amounted to near £600', and the overall total Boyle provided during the remainder of his life – as also through his will – was almost certainly substantial.[34]

It was in parallel with this that the most celebrated example of Boyle's philanthropic concern with Ireland occurred: his patronage of the printing of the Bible in Irish, which was also, of course, what brought him into contact with Narcissus Marsh, and it is on this that I next want to dwell. The earliest of the numerous letters discussing the subject dates from December 1678: this is a response to a missive from Boyle of 12 December that year, now lost, to the former Jesuit Andrew Sall, whose conversion to Anglicanism in 1674 had represented something of a propaganda coup for the Protestant cause, and about whose role in the affair I will say more shortly. Why did Boyle get involved, and why did this episode come to the fore at the time it did?

When giving a retrospective account of it to the Scottish cleric James Kirkwood, Boyle wrote as follows:

---

[32] For such views, see esp. J.R. Jacob, *Robert Boyle and the English Revolution* (New York, 1977).

[33] *Correspondence*, vol. 4, pp. 320ff.; see esp. the document printed in ibid., pp. 366–71.

[34] Birch, 'Life of Boyle', in his edition of *The Works of Robert Boyle* (2nd edn, 6 vols, London, 1772), vol. 1, p. cxxxix. Cf. ibid, pp. cxxxix–cxl; *Correspondence*, vol. 6, pp. 88–9, 91–2, 157–8, 200–201; Maddison, *Life of Boyle*, pp. 273–4.

at first my Aime was, as I thought it became me, to do some service to the Countrey wherein I was born, & have some little Estate, tho I have lived a great Stranger to it: and especially to contribute to the Conversion & the Instruction of those Irish Natives that are most of them of the Romish Religion.[35]

Though written with hindsight, this rings true as far as it goes. By way of background, one can adduce the concern to make proper use of Boyle's Irish revenues seen in the solicitousness for the ministers in the impropriated parishes that I have just referred to. This could easily also explain why the project arose at the time it did, but there are certain other clues which deserve scrutiny.

One piece of evidence adduced by R.E.W. Maddison in his study of Boyle and the Irish Bible is a copy in the hand of Boyle's principal amanuensis in his later years, Robin Bacon, of a document dated August 1675 which presents various strategies for the improvement of Ireland over the next nine years.[36] In fact, although Maddison was not aware of this, the document is by Sir William Petty, and various copies of it survive among Petty's manuscripts, now in the British Library.[37] Most of the suggestions are for such projects as 'That some where in the west of Ireland, There be a free port, and place for building, repairing, victualling & equipping Ships of Warr, equivalent to the tenth part of the Navy of England, with Dock-yard, store-house, and Rope-yard', or 'That a Stock be found whereby to double the present trade & exportation; at under 8 per cent Interest, and 5 per cent Exchange between England & Ireland with perpetual Rules for Equalizeing & regulateing the several species of Mony and Coynes'. However, no. 4 is: 'That the Bible and Common Prayer-book be translated into Irish and printed in the vulgar character and that fit Persons be salariated to preach catechize and officiate in Irish in every part of that kingdom weekly'.

Though it is quite likely that Boyle obtained this document soon after it was written,[38] there is no reason to think it acted as a stimulus to him in this matter, as Maddison implied by citing it in this context. In fact, it is one of a number of miscellaneous papers relating to Irish affairs scattered through the Boyle archive, evidently there because of their relevance to his Irish concerns, and not least his landholdings, beyond which it is difficult to say very much about them.[39] In

---

[35]   *Correspondence*, vol. 6, pp. 251–2.

[36]   BP 40, fols 143–4. R.E.W. Maddison, 'Robert Boyle and the Irish Bible', *Bulletin of the John Rylands Library*, 41 (1958), 81–101, on pp. 81–2.

[37]   BL Add. MS 72789, fols. 84–97. A further copy, owned by Southwell, is in Add. MS 72852, fols 100–101. I am indebted to Dr Frances Harris for these references.

[38]   Of the various versions of the text in Add. MS 72879 (and 72852), the BP 40 text is most similar to fols 88–9, which apparently dates from September 1675, as against subsequent versions with additional material.

[39]   For example, BP 40, fols 76–9, 93–9, 138–42; RS MS 186, fols 187–8; MS 190, fol. 144v. See *Boyle Papers*, pp. 486–9, 541, 563.

particular, the rather manipulative tone, though typical of Petty, does not really ring true to the conscience-ridden view of Ireland on Boyle's part that I have already sketched.

A more plausible background is provided by the last part of Boyle's explanation to Kirkwood, which was echoed by Narcissus Marsh when expounding the Bible project to Archbishop William Sancroft in August 1682, as being 'for the instructing the poor deluded blind Natives of this Kingdome, who are now shut up in a miserable darkness, & differ but little from Heathens, save that they bear the name of Christians'.[40] Marsh's phraseology, even more than Boyle's own, establishes a parallel between the Irish enterprise and the broader evangelical efforts throughout the world that so preoccupied Boyle. I have already referred to Boyle's Governorship of the New England Company, one of the chief activities of which was to proselytize the native population by publishing devotional texts in their native language, Algonquin. Interestingly, Boyle had recounted the Company's achievement in this regard in a letter of 5 March 1677 to Robert Thompson, a member of the court of the East India Company, in connection with the suggestion that that company might take on a similar role in the Far East; 1677 also saw the publication of the Gospels in Malay at Boyle's expense, under the auspices of Thomas Hyde, assisted by Thomas Marshall.[41] It seems fairly clear that it is in the context of this broader evangelistic mission that the Irish Bible project is to be seen.

As already noted, the earliest extant letter relating to the project is that to Boyle from Andrew Sall dated 17 December 1678. It was written from Oxford, and this provides a further important part of the background to Boyle's Irish links at this time. Boyle himself had moved from Oxford to London in 1668, but he continued to have close contacts with Oxford, and particularly with a group of eminent scholars there with whom he shared various concerns – the interest in biblical scholarship to which Ussher had introduced him in the 1640s, an equally deep interest in various aspects of natural philosophy, and a commitment to evangelical work. Both Hyde and Marshall, whom I mentioned in relation to the Malay gospels, were Oxford dons, as was Edward Pococke, whose translation of Grotius' *Of the Truth of the Christian Religion* into Arabic, published in 1660, had been sponsored by Boyle.[42]

Boyle also had links with other men at Oxford whose names will recur later in this chapter, including the Savilian Professors of Astronomy and Geometry Edward Bernard and John Wallis, while another influential Oxford figure who shared various of Boyle's interests, although his High Churchmanship was not

---

[40]   Bodleian MS Tanner 35, fol. 74.
[41]   *Correspondence*, vol. 4, pp. 436–8, and ibid., pp. 426ff., passim.
[42]   *Between God and Science*, p. 123; *Correspondence*, vol. 1, pp. 382, 426–8, 449–51.

to Boyle's taste, was John Fell, Dean of Christ Church and Bishop of Oxford.[43] Fell was responsible for one of the most significant intellectual initiatives at Oxford during this period, the promotion of an active programme of scholarly publishing, especially of patristic and early Christian texts, at the University Press there. Of the products of the Press, perhaps the most notable was the *Synodicon*, a collection of the canons of the Eastern Church edited by the divine and later Bishop of St Asaph William Beveridge, in which Narcissus Marsh had served a significant editorial role.[44] Fell also played a part in the attempt to involve the East India Company in evangelical activity in 1677 to which I have just referred; the idea was that Oxford might act as a nursery for missionaries, 'furnishd not only with the Arabick Tongue but, if it were desired, with Arithmetick and other parts of the Mathematicks & other Qualifications fit to recommend them, & make them appear more considerable & grow more usefull in those parts'.[45] In addition, Fell was more or less directly responsible for the arrival in Ireland of most of Boyle's main contacts in the Irish Bible project: Andrew Sall, for whom Fell provided lodgings at Christ Church during his time in Oxford, and then Narcissus Marsh and Robert Huntington, both of them Oxford-educated protégés of Fell who successively became Provost of Trinity College, Dublin, on Fell's recommendation to the Duke of Ormonde, Vice-Chancellor of Oxford as well as Lord Lieutenant of Ireland.[46]

The first of these, the Jesuit convert Andrew Sall, explained in his initial letter to Boyle that his 'present work, comended to me by our chief prelates', was to update his 'course of Philosophie controversiall and morall' and his 'fundamentall tracts of practicall theologie' by citing 'the more creditable authors of this age' and using 'a more cleer and bræf method'; this was evidently for use in Ireland, where he was to return in 1680.[47] His comments make it perfectly clear that the 'designe of reprinting the new testament in Irish, and how it may conduce to the conversion of those miserably deluded souls' was the initiative of Boyle, but at Boyle's behest Sall now took up the project, and it was he who, following his return to Ireland, was mainly responsible for canvassing support

---

43   For Boyle's relations with Fell, see *Boyle by Himself and His Friends*, pp. lxx, 33–4, 71.

44   See Harry Carter, *A History of the Oxford University Press, Volume 1: To the Year 1780* (Oxford, 1975), chs 5–10 (esp. pp. 47–9 on the *Synodicon*); Michael Hunter, *Science and the Shape of Orthodoxy* (Woodbridge, 1995), ch. 10. For Marsh's assistance with the *Synodicon*, see 'Archbishop Marsh's Diary', *British Magazine*, 28 (1845), 17–26, 115–32, on p. 21.

45   *Correspondence*, vol. 4, p. 437. Cf. *Boyle by Himself and his Friends*, pp. 33–4.

46   For Sall, see *ODNB*. For Marsh, see 'Archbishop Marsh's Diary', p. 22. For Huntington, see Thomas Smith, 'De vita, studii, peregrinationibus, & obitu ... D. Roberti Huntingtoni', in Huntington, *Epistolæ* (London, 1704), pp. xxvi–xxvii; translated as 'The Life and Travels of the Right Rev. and Learned Dr Robert Huntington', *Gentleman's Magazine*, 95 (1825), 11–15, 115–19, 218–21, on p. 119.

47   *Correspondence*, vol. 5, pp. 133–6.

for the project there. He also arranged for the text of the New Testament to be prepared for the press, along with the catechism that accompanied it, both of them in fact based on earlier published exemplars. Meanwhile, in London, Boyle liaised with typecasters and printers, even having a new font of Irish type made by the leading typographer in Restoration London, Joseph Moxon, perhaps to a design by Sall.[48] This was necessary since the type used for the former biblical and other texts printed in Irish could not be traced; the type produced at Boyle's behest was to remain the standard type for Irish books printed in Britain for over a century.[49]

Sall was responsible for the preface to the edition, in some ways a slightly curious text with a curious history. It appears that Boyle had suggested that the preface might be based on the preface to the translation of the New Testament into French prepared by the Jansenists and published in 1667. Evidently, he felt that this offered a cogent argument for translating the scripture into the vernacular; he may also have felt that the fact that it was written by Roman Catholics might have increased its appropriateness (and it is worth noting that, although anxious to save the souls of the native Irish as of the heathen elsewhere in the world, Boyle seems to have lacked the fanatical anti-Catholicism of some of those involved in this project).[50] However, Sall and others who read the piece thought it too overtly anti-Protestant to be appropriate, and Boyle had to admit

---

[48]   Ibid., vol. 5, passim; Maddison, 'Boyle and the Irish Bible', esp. pp. 83–5 (which reproduces a memorandum between Boyle and the printer, Robert Everingham); Joseph Moxon, *Mechanick Exercises on the Whole Art of Printing*, ed. Herbert Davis and Harry Carter (London, 1958), pp. xxxviii, 366–7, 370; E.W. Lynam, 'The Irish Character in Print', *The Library*, 4 (4) (1924), 286–325, on p. 301. See also Talbot Baines Read, *A History of the Old English Letter Foundries*, rev. ed. by A.F. Johnson (London, 1952), p. 177; Edward Rowe Mores, *A Dissertation upon English Typographical Founders and Foundries* (1778), ed. Harry Carter and Christopher Ricks (Oxford Bibliographical Society, NS 9, 1961), pp. lxxvi, 32, 98.

[49]   Lynam, 'Irish Character', p. 301. For the fate of the earlier type, see Maddison, 'Boyle and the Irish Bible', p. 84, and Barnard, *Cromwellian Ireland*, p. 179, n. 200 (though the link with the events at Athenry is unclear). Certain punches, formerly in the possession of Stephenson Blake Ltd of Sheffield, are now in the Type Museum, London.

[50]   See, for instance, Boyle's comments in a lost letter from him quoted in a letter from Robert Huntington to Anthony Dopping, Bishop of Meath, 10 October 1685, in which he expressed his preference for 'mollifying 2 or 3 expressions, that may seem somewhat severe to the Romanists' in the preface that Dopping had written for the Irish Old Testament (see below; the preface was not published and does not survive), though he deferred to the views of those on the spot: Armagh Public Library, Dopping Collection, no. 46. This forms part of a group of letters that came to light too late to be included in the printed edition of Boyle, *Correspondence*. It is included in the supplement to the electronic edition published by InteLex, Charlottesville, Virginia (2003), and is also available on the Boyle website, http://www.bbk.ac.uk/boyle/researchers/works/correspondence/Corresp.%20Suppl.%20rev.pdf.

that, in recommending it, he had been relying more on the opinion of others to whom he had lent the original than his own hasty appraisal of it.[51] Instead, Sall went on to write a rather different preface, addressed 'To the Christian People of Ireland', which appeared at the start of the volume in both English and Irish.[52] This invoked patristic authority for the legitimacy and importance of reading the scripture in the vernacular (this appeal to the Fathers was typical of the school of churchmanship associated with Fell and others at Oxford). It also gave a fulsome account of how:

> God has raised up the generous Spirit of *Robert Boyle* Esq; Son to the Right Honourable *Richard* Earl of *Cork*, Lord High Treasurer of *Ireland*, renowned for his Piety and Learning, who hath caused the same Book of the *New Testament* to be Reprinted at his proper Cost; And as well for that purpose as for Printing the *Old Testament*, and what other Pious Books shall be thought convenient to be published in the *Irish* Tongue, has caused a New Set of fair *Irish* Characters to be Cast in *London*, and an able Printer to be instructed in the way of Printing this Language.

In addition, in an evident attempt to appeal to a native readership, the preface made a rather strange and awkward allusion to Christ's miracles as exceeding those of the heroes of legends and romances.[53]

Also rather awkward was the way Sall attempted to address one of the chief difficulties of the project, the fact that many opposed the publication of the Bible or any other text in Irish, on the grounds that it would perpetuate a language which they would rather see extirpated. Sall wrote: 'notwithstanding all the wise Statutes and Endeavours used to bring this whole Nation to a Knowledge of the *English* Tongue, Experience shews, it could not be effected, too many being unable to give such Teaching to their Children, or get it for themselves'.[54] Yet this was precisely what many would have denied, as was cogently expressed in a letter to Boyle by the other figure chiefly involved in the early stages of the project, Henry Jones, Bishop of Meath until his death in 1682, who had in fact been connected with attempts to publish the Bible in Irish earlier in the seventeenth century. Having discussed the matter with Irish parliamentarians, Jones 'found it almost a principle in theire Politiques, to suppresse that Language utterly, rather then in so publique a way to countenance it'.[55] This robust opposition

---

[51] *Correspondence*, vol. 5, pp. 220, 265–9, 274–6.

[52] 'To the Christian People of Ireland. The Preface', in *Tiomna Nuadh* (London, 1681), sigs a1–b4.

[53] Ibid., sigs a1v–2, a3.

[54] Ibid., sig. b1.

[55] *Correspondence*, vol. 5, p. 208. For Jones's links with the earlier initiatives, see Barnard, 'Protestants and the Irish Language', p. 248; see also his 'Crises of Identity among

was to dog the project, as it had its predecessor half a century before, when the translation of the Old Testament prepared by Bishop William Bedell had never been published.[56] Yet Boyle and his clerical supporters seem to have been entirely unaffected by any such reservations, no doubt on the grounds that, in the words of a commentator on the earlier polarisation between the supporters of the provision of such texts and their opponents, 'the reasons of the former were drawne from the principles of theology, and the good of soules, of the latter, from politicks & maxims of state'.[57]

It seems to have been Henry Jones who initially suggested that the new edition of the New Testament in Irish might be accompanied by the publication for the first time of an Irish translation of the Old. For this, he proferred the translation made by Bishop Bedell in the 1630s, the manuscript of which he had in his possession, and the preparation of this for the press was put in hand.[58] Initially, this took place under the auspices of Andrew Sall, but he died suddenly on 5 April 1682, and this brought Narcissus Marsh to the fore as the chief Irish agent of the project.[59]

The result was a lengthy exchange of letters between Marsh and Boyle, and, in contrast to the state of affairs with Sall, where we have only Sall's letters to Boyle rather than vice versa, in this case the bulk of the correspondence on both sides survives. From it, one gains some intriguing vignettes both of Boyle and Marsh. For instance, on one occasion Boyle had to apologize for 'a Disaster' that befell one of Marsh's letters to him, which:

> being brought yesterday from the post, not directly to me, but to a Servant that was then busyed about the fire, to make a Chymical Experiment I had orderd him to attend in my absence, he haveing laid it by for a while, a kindled Coal unluckily

---

Irish Protestants', *Past and Present*, 127 (1990), 39–83, on pp. 57–8.

[56] For Bedell's translation, see Gilbert Burnet, *The Life of William Bedell* (London, 1685), pp. 117ff., and E.S. Shuckburgh (ed.), *Two Biographies of William Bedell* (Cambridge, 1902), pp. 55–6, 131ff.

[57] *Correspondence*, vol. 6, pp. 451–2, a letter of 14 December 1685 to Huntington from Anthony Dopping.

[58] See ibid., vol. 5, p. 208 and passim. Jones had divulged his possession of Bedell's MS translation in the dedicatory epistle to his *Sermon of Antichrist* (Dublin, 1676), sig. A2v, where he stated that he wished 'it were for such a publique good, printed and published': but the time lag between this and its inclusion in the project suggests that Boyle was not aware of this.

[59] See Maddison, 'Boyle and the Irish Bible', pp. 85, 93–7. At this point, Maddison was unaware of the whereabouts of the Marsh letters: see ibid., p. 97. For their discovery, see *Correspondence*, vol. 1, p. xxiv.

lighted on the Letter, and burnt it quite thorow in that part that contain (as I conjectur'd) some of the most important passages of it.[60]

More significantly, it is clear that there was a real rapport between the two men, and it is worth commenting on this here. It should be apparent from what I have already said about Boyle's links with Ussher and with the Oxford scholars with whom he worked, such as Thomas Hyde and Thomas Marshall, that he had a deep respect for scholarship. In the case of Marsh, his scholarly concerns are demonstrated both by his correspondence and by his remarkable collection of manuscripts, now in the Bodleian Library.[61] Equally, Marsh clearly had real respect for Boyle's achievement as a natural philosopher. In a letter of introduction he wrote to Boyle on behalf of the young scholar Thomas Molyneux when he set out for England in 1683, he explained: 'Nor can any one pretend to an acquaintance with the ingenious world abroad, who besides your Books has not some knowledg of your person and your admireable contrivances for the emprovement of Experimentall philosophy.'[62] Marsh was himself a natural philosopher, having written an innovative piece on the properties of vibrating strings which was published in Robert Plot's *Natural History of Oxfordshire* (1677), while more will be heard about his scientific interests and their intersection with Boyle's own concerns shortly.[63] Perhaps above all, Marsh clearly shared Boyle's deep faith, as is apparent in Marsh's rather revealing 'Diary', which echoes some of the introspective documents associated with Boyle I have already referred to in its anxious providentialism, its record of deliverances and monitory dreams and its constant soul-searching. For all Marsh's efficient management of the affairs for which he was responsible, his diary reveals 'a truly

---

[60] *Correspondence*, vol. 5, p. 316, and vols 5–6, passim.

[61] In addition to the letters to Bernard cited below, note also those to Arthur Charlett, 1696–1709, in Bodleian MS Ballard 8, fols 3–18, and those to Thomas Smith, 1697–1709, in Bodleian MS Smith 52, fols 51–158. For an overview of Marsh's MSS, see Falconer Madan, H.H.E. Craster, R.W. Hunt and P.D. Record, *A Summary Catalogue of Western Manuscripts in the Bodleian Library at Oxford* (7 vols, Oxford, 1895–1953), vol. 3, pp. 45–61.

[62] *Correspondence*, vol. 5, pp. 406–7.

[63] Robert Plot, *The Natural History of Oxfordshire* (2nd edn, Oxford, 1705), pp. 293–305. On the background to this, see Penelope Gouk, *Music, Science and Natural Magic in Seventeenth-century England* (New Haven, CT and London, 1999), pp. 50, 53–4. Boyle's interest in related topics is shown not least by his possession of the MS of John Birchensa's 'Compendious Discourse of the Principles of the Practicall & Mathematicall Partes of Music', BP 41, fols 1–21, stated to have been written 'for the use of the Honourable Robert Boyle Esq.'; this is followed by a holograph letter from John Wallis to Henry Oldenburg on related topics.

experimental piety', in the words of an early nineteenth-century commentator, which would have rung a sympathetic chord with Boyle.[64]

Hence Boyle and Marsh seem to have formed an effective team in the Bible project, despite the problems that emerged and that are apparent from their letters. Apart from the undercurrent of hostility to the entire project, which if anything intensified as the 1680s progressed, there were also practical problems concerning the preparation of the copy. Whereas the printing of the New Testament had been relatively straightforward, delayed only by the fact that the London printing houses were sometimes too preoccupied by pamphlets relating to the Exclusion Crisis to undertake work of this kind,[65] with the Old Testament a whole series of complications arose. For one thing, the manuscript of Bedell's text proved to be 'a confused heap pittiefully defaced and broken': this meant that it had to be corrected and re-transcribed to make it comprehensible to the printer, causing delay and additional expense.[66] In an attempt to recoup the costs involved, attempts were made to gain subscriptions in support of the enterprise, on the model of printed proposals used to raise funds for the Welsh Bible published in 1677, though with little success.[67] Indeed, the response even of various bishops of the Church of Ireland to this request for help were rather negative, and in one case directly critical of Boyle.[68] There was also disagreement about the degree of revision of Bedell's original text that was appropriate.[69] In all these matters, and not least the latter, Marsh played a central role; he explained in a much later letter to the antiquary Edward Lhuyd how revision occurred:

> in some places, where the Translation did not fully express the sence of the Hebrew Text; which alteration was made upon due consideration, & with the advice of men well skill'd in both languages, which cost me no small pains, & more than a years hard labour at times.[70]

---

[64] 'Some Passages in the Life of Narcissus Marsh', *Christian Examiner*, 11 (1831), 645–50, on p. 645. For Marsh's diary, see above, n. 44.

[65] *Correspondence*, vol. 5, pp. 249–50.

[66] Ibid., vol. 5, p. 279. Cf. ibid., pp. 337–9, 347, 377–8, 387–9; vol. 6, pp. 5, 106. The original MS now survives in Marsh's Library.

[67] Ibid., vol. 5, esp. pp. 288, 356–8, 387.

[68] For the difficulties encountered in raising these subscriptions, see various letters in the Dopping Collection at Armagh (above, n. 50), esp. nos 24–6, 28–33, quoted at length in Betsey Taylor Fitzsimon, 'Conversion, the Bible and the Irish Language: The Correspondence of Lady Ranelagh and Bishop Dopping', in Michael Brown, C.I. McGrath and T.P. Power (eds), *Converts and Conversion in Ireland, 1650–1850* (Dublin, 2005), pp. 157–82, on pp. 173–5.

[69] *Correspondence*, esp. vol. 5, p. 387, vol. 6, p. 106.

[70] Bodleian MS Ashmole 1816, fol. 323v: Marsh to Lhuyd, 4 October 1707. A further later account of the project appears in Bodleian MS Smith 52, fols 121–8, Marsh to Smith,

Boyle was also supportive of other projects with which Marsh was involved during his time at Trinity. He clearly approved of Marsh's initiative in setting up an Irish lectureship in the college, despite the fact that this, too, was frowned upon by some, so that in 1682 Marsh felt obliged to ask the Archbishop of Canterbury whether he should continue with it.[71] Equally important, Boyle supported and offered to help with the printing of the grammar of the Irish language on which Marsh was working at this time – an interesting and perhaps surprising initiative on the part of an erudite linguist like Marsh. In the end, this failed to come to fruition, and the manuscript was apparently lost in the upheavals following the Revolution of 1688–89, but the grammar is discussed in many letters between the two men.[72] Boyle evidently approved of it, not least because of the potential it offered for evangelization; he offered advice on various facets of the project, while also expressing his satisfaction at the prospect that the type he had had cast should be used to print it.[73]

What is equally interesting is that, in parallel with the exchanges relating to the Bible project, one also sees a liaison developing from 1680 onwards between Marsh, Boyle and Boyle's Oxford contacts concerning natural philosophy, hence constituting an important element in the background to the foundation of the Dublin Philosophical Society in 1683 that has not hitherto received the attention it deserves. In particular, this concerned the comets that awed and intrigued the whole of Europe in 1680–81 and 1682.

Marsh made detailed observations of the comet of 1680–81, at intervals of as little as 15 minutes at times when it was fully visible in Dublin, though he was characteristically diffident about these records, describing them as:

> imperfect & unconstant, partly for want of good Instruments (my Telescope being the best I had to confide in) partly through the variety of the weather; but chiefly because the College standing East of the City, I was forct to observe it alwaies <over the City & so> through the smoke of the town, which many times render'd it invisible here, when 3 miles off the place it might well be seen.

He was also typically self-effacing over matters that he wished he had 'diligently taken notice of', but had failed to.[74]

---

19 January 1706 (printed in *The Christian Examiner*, n.s. 2 (1833), 761–72), correcting the account given in Smith's 'Vita ... D. Roberti Huntingtoni' (for which see above, n. 46), pp. xxviii–xxix.

[71] Bodleian MS Tanner 35, fol. 74. See also *Correspondence*, vol. 5, pp. 253–4, 272; *Christian Examiner* (see above, n. 70), pp. 768–9.

[72] See the Marsh–Lhuyd letter referred to above, n. 70.

[73] *Correspondence*, vol. 5, pp. 281, 286, 290, 315, 387, 398, 405; vol. 6, pp. 125, 137–8, 174–5.

[74] Bodleian MS Smith 45, p. 27: before 'variety', 'incon' is deleted.

Initially, these observations were sent to Boyle by Henry Jones as an appendage to a letter concerning the Irish Bible,[75] but in parallel with this – delayed by his commitments as Provost of Trinity – Marsh completed a detailed letter of his own, which he sent to the Savilian Professor of Astronomy at Oxford, Edward Bernard. Marsh had earlier corresponded with Bernard concerning an eclipse in 1679, which he had been warned to look out for by Edmond Halley, while in 1680 he had sent Bernard an elaborate list of the Greek manuscripts at Trinity, although he was slightly apologetic about their quality; he also sent information for two other Oxford scholars, Hyde and Pococke.[76] In the lengthy letter he now wrote, begun in April 1681 and finished in May, he not only presented his observations but explained the various theories he had sought to test through his empirical work, including the extent of parallax, the nature of the comet's tail, and its position in relation to the fixed stars and planets. In addition, he outlined his 'Hypothesis of Comets', by which he believed 'all the phænomena of this Comet may be solv'd', illustrating it with an elaborate diagram (Plate 2). In his view, this and other comets were the remains of *faculae* 'cast of from the sun, by it's revolution on it's own Axis', 'whose train was nothing but the Sun-Beams darted through it's body, like raies through a burning glass, or rather through a glass ball fill'd with water'.[77]

Marsh's letter, though addressed to Bernard, was clearly perceived as being a general missive 'to us at Oxford'; among others, it was seen by the astronomer royal, John Flamsteed, whose notes on Marsh's data concerning the comet survive along with similar records of observations of the comet from other sources.[78] A reply to it was composed, not by Bernard, but by the Savilian Professor of Geometry John Wallis, a long-term contact of Boyle's: it was to Boyle that Wallis had addressed a seminal paper on the movement of the tides in 1666, while (as he divulged in this letter) it was also to Boyle that he initially propounded, 'privately, & he would needs have me speak it out', his theory that there was a link between the fact that 'about the year 1650 & before, spots were so frequently seen on the sun', whereas 'since the year 1650, we had seen a great many comets (more than in many years before)'.[79] In his letter to Marsh dated 16 June 1681, Wallis offered his own elaborate theory concerning the mutual relationship of the comet with the Earth and the Sun,

---

75   *Correspondence*, vol. 5, pp. 254–5. Cf. ibid., p. 264.

76   Bodleian MS Smith 45, pp. 15–16, 19–20, 27–30, 25–6 (the letter of 9 April/14 May 1681 is bound out of order, with the final leaf first).

77   Bodleian MS Smith 45, pp. 27–30, 25: after 'rather', 'like' is deleted.

78   Bodleian MS Don d. 45, fol. 171b; Cambridge University Library, Royal Greenwich Observatory MS 1/41, fol. 23v. See E.G. Forbes, Lesley Murdin and Frances Willmoth (eds), *The Correspondence of John Flamsteed* (3 vols, Bristol, 1995–2001), vol. 1, pp. 794, 811. I am grateful to Frances Willmoth for her help in this connection.

79   See *Correspondence*, vol. 3, pp. 141–56; Bodleian MS Don d. 45, fols. 171a–b.

commenting only obliquely on Marsh's theory; though he professed to have 'a very favourable opinion' of this, he claimed that it overlapped with the view of the early seventeenth-century astronomer Jeremy Horrocks, whose papers had been published at the behest of the Royal Society in 1673.[80] Interestingly, Wallis dispatched this letter not directly to Marsh, but via the Royal Society, inviting the then Secretary, Thomas Gale, to show it to anyone who might be interested, which may have included Flamsteed and Boyle.[81]

Flamsteed was also stimulated to comment on Marsh's letter in a letter to Bernard of 8 September 1681: here, as in his notes, his concern was with correlating Marsh's data about the size and course of the comet with that available from other sources.[82] Meanwhile, Marsh himself had responded to Wallis's missive with a further letter dated 6 August that year, in which he was enthusiastic about Wallis's theory, though expressing some reservations (thus revealing no sense of inferiority to an acknowledged master in the field); he also recorded some second thoughts about his own hypothesis.[83] This stimulated two further long letters from Wallis, probably also shown to others than Marsh, and Wallis's perception of the significance of the whole exchange is shown by the fact that he devoted a whole section of his commonplace book to it, entitled 'A Comet. 1680'.[84]

In the case of the comet of 1682 – Halley's comet – Marsh initially gave Boyle notice of it in a letter which is unfortunately not extant; subsequently, he reported that he had 'been so unhappy, as not to be able to make any observations on the late Comet, nor to see it more than twice; when it's head appear'd to the naked eie much bigger, & it's tail much shorter than that of the last'. Boyle's response shows his own interest in the comet: he noted how, having observed it with a glass, 'I judgd it somewhat greater in body than the last; but

---

[80]   Ibid.

[81]   Ibid., fols 171a–172. For Wallis's letter to Gale, see also Royal Society Copy Letter Book, vol. 8, pp. 236–8. See also *Correspondence of Flamsteed*, vol. 1, pp. 795–6; Birch, *Royal Society*, vol. 4, p. 93. On Flamsteed, see below, n. 82.

[82]   See *Correspondence of Flamsteed*, vol. 1, pp. 809–11. For the possibility that Flamsteed had already commented on Marsh's letter in a lost letter to Bernard dating from earlier in 1681, see ibid., p. 794.

[83]   Bodleian MS Don d. 45, fols 172v–173.

[84]   Ibid., fols 173v–175 and 165–75, passim. Copies of Wallis's letters also survive in Royal Irish Academy, Dublin, MS 12/D/34, pp. 1–18, a manuscript compiled by an associate of Marsh's, probably after his death: see Edward McPharland, 'Building Marsh's Library: The Architecture of Marsh's in its European Context', in Muriel McCarthy and Ann Simmons (eds), *The Making of Marsh's Library* (Dublin, 2004), pp. 41–50, on p. 48. It is perhaps worth noting that the next items Wallis copied into his commonplace book were a pair of letters in which he gave some quite technical mathematical advice to Boyle: *Correspondence*, vol. 6, pp. 115–16, 121–2.

of a, comparatively, very short taile'. However, he deferred to 'our Mathematical Virtuosi' for a fuller account of it, thus echoing Marsh's comment that 'the virtuosi in London, who are alwaies intent on what's rare, I presume have not let it escape their accurate observations'.[85]

Returning to Marsh's initial letter to Bernard, this is equally interesting for the evidence it provides of the state of Dublin science in 1680–81. Marsh records how Petty as well as he made observations of the comet that appeared at that point, while in making his own – in which he used a 16-foot telescope as well as a 3-foot one – Marsh was assisted by 'some others of this College, my assistants'.[86] He also explains how he illustrated one aspect of his theory of comets by showing how light was refracted through a glass jar filled with water, by which 'I represented it to the satisfaction of all the beholders ... (when the Comet on the suddain disappear'd, & so frustrated the Expectations of severall, who came to me on purpose to observ it)'.[87]

Marsh also reported accounts of 'a fiery Meteor flying swiftly through the air very low over Dublin, which made a hissing noise, very audible as it pass'd along'. ''Twas very big & bright,' he told Bernard, adding: 'I saw it not my self, & the reports of those, who did, are so various, that I shall say no more of it.' In addition, he blamed his delay in completing his initial letter to Bernard in part on the distraction of 'not a few letters extorted from me on this subject, by men more inquisitive than knowing in this Kingdome'.[88] The sightings of 1680–81 may also provide a background to the Dublin Philosophical Society in terms of the politics of Ireland at the time, as revealed by a further related text, a printed pamphlet entitled *A Judgement of the Comet. Which became first Generally Visible To Us in Dublin December XIII About 15 Minutes before 5 in the Evening Anno Dom. 1680* by 'a Person of Quality', usually identified as Edward Wetenhall, Bishop of Cork and Ross and later a member of the Dublin Philosophical Society.[89] For 'The Stationer to the Reader' averred how 'the sole Occasion of his writing of it was the strange consternations of many People at that time'; apprehensions that the comet might presage a new massacre of Protestants meant that 'In some places several Families would get together in one house by night, and a certain number watch while the rest slept.'[90] Wetenhall commented on the comet as 'a Rational person and considering Christian', arguing that the production of comets was 'nothing supernatural' and that they had no significance other than 'to beget in us a deeper belief and admiration of the great

---

[85]   *Correspondence*, vol. 5, pp. 337, 348. Cf. ibid., pp. 340, 352.

[86]   Bodleian MS Smith 45, pp. 27, 30.

[87]   Ibid, pp. 27, 30, 25.

[88]   Ibid., pp. 27–8: after 'Dublin', 'Nov the' [?] is deleted.

[89]   K.T. Hoppen, *The Common Scientist in the Seventeenth Century: A Study of the Dublin Philosophical Society, 1683–1708* (London, 1970), pp. 45–6.

[90]   Wetenhall, *A Judgement* (Dublin, 1682), sig. a1.

God of Heaven and Earth'.[91] In other words, he wanted to reassure people by invoking 'science', and the new society clearly helped in that regard – though I should disavow any reductionist intention in making this passing suggestion, which I know some would have used as the basis for a revisionist, constructivist account of the origins of Irish science.

In parallel with these developments, William Molyneux's observations of the lunar eclipse at Dublin in August 1681 led to the establishment of a correspondence between him and John Flamsteed which was to continue for many years, and this, too, forms a significant part of the background to the foundation of the Dublin society.[92] A further initiative on Molyneux's part was his agreement to provide the Irish section of the *Atlas* being prepared by the London publisher Moses Pitt, with the encouragement of John Fell and others, though this was ultimately to prove abortive.[93] The first reference to the Dublin society in Marsh's correspondence is in this connection: he told 'My Lord' – probably Michael Boyle, now Archbishop of Armagh – in a letter of 18 May 1682 that 'We are now (a club of us, who meet every week in the College) upon the design of giving an account of Ireland to be printed in the New Atlas'.[94] Moreover, in this connection, Molyneux had a set of 'queries' printed, which clearly derive from similar series issued by figures like Robert Plot in connection with his county natural histories in the 1670s, themselves based on similar questionnaires devised by Boyle and published in the Royal Society's *Philosophical Transactions* in the 1660s.[95]

From such initiatives, the Dublin Philosophical Society sprang, coming formally into being in 1683, the same year in which comparable institutions

---

[91]   Ibid., contents list, pp. 1, 50 and passim.

[92]   *Correspondence of Flamsteed*, vol. 1, pp. 807 ff. and passim. See also Hoppen, *Common Scientist*, pp. 23, 89, 113, and J.G. Simms, *William Molyneux of Dublin* (Blackrock, 1988), pp. 23ff. It is perhaps worth noting here that Molyneux further exemplifies the overlap between natural philosophy and the Bible project in Boyle's Irish connections, since he was the brother-in-law of Anthony Dopping, and it was he who told Dopping about Boyle's increased contribution to the costs of printing the Old Testament in a letter of 25 August 1682: see Armagh Public Library, Dopping Collection, no. 27, and Boyle, *Correspondence*, vol. 5, pp. 314–15.

[93]   See E.G.R. Taylor, 'The English Atlas of Moses Pitt', *Geographical Journal*, 95 (1940), 292–9.

[94]   Bodleian MS Rawl. Letters 45, no. 14. See also Hoppen, *Common Scientist*, pp. 21–2, and Simms, *William Molyneux*, ch. 3.

[95]   Hoppen, *Common Scientist*, pp. 21–2, 200–201; see Chapter 3 in this volume. I should note that the alternative potential ancestry for Molyneux's queries in Hartlib's *Interrogatory*, suggested, for example, in Barnard, *Cromwellian Ireland*, p. 237, is rendered less likely by the fact that Molyneux's queries follow the thematic arrangement of Boyle and Plot rather than the alphabetical arrangement of Boate: see above, p. 70 n. 60.

were founded at Oxford and at Boston, Massachusetts. I need say little about the society and its activities, which have been definitively studied by K.T. Hoppen;[96] what is significant here is that, from 1683 onwards, a regular pattern of exchange developed between Boyle and the virtuosi who made up the Dublin society. Clearly Boyle was in frequent contact with Petty, whom the young Thomas Molyneux found at Boyle's lodgings when he called on the great man on 22 May 1683, presumably with the letter of introduction from Marsh that has already been referred to. Molyneux went on to give his brother William a memorable account of the ageing Boyle himself: 'he stutters, though not much; speaks very slow, and with many circumlocutions, just as he writes.'[97] Another visitor was Edward Wetenhall, who visited Boyle and viewed his laboratory in 1683 and who later wrote: 'I can scarce recollect any time, which I now account better spent while I was in London, than that which I employed in visiting you.'[98]

Most significant were the links Boyle now developed with the physician and naturalist Allen Mullen, one of the most active members of the Dublin Philosophical Society in its earliest years, who in 1681 had achieved a certain notoriety by dissecting an elephant which accidentally died while on show in Dublin. Mullen published an account of this in 1682, accompanied by an essay on the comparative anatomy of animals' eyes which was fulsomely addressed to Boyle.[99] Two long letters survive from Mullen to Boyle, dated 2 December 1682 and 26 February 1686, in which he expatiates at length on their shared interests, in the second commenting on Boyle's recently published *Reconcileableness of Specifick Medicines to the Corpuscular Philosophy* (1685); a further letter dated September 1685 is known once to have existed, but is now lost.[100] In 1688, Mullen moved to London and was active in scientific circles there, as is recorded both in the minutes of the Royal Society and in Robert Hooke's diary.[101] He also actually worked with Boyle, carrying out an experiment on a dog at Boyle's house

---

[96] In addition to Hoppen, *Common Scientist*, see the valuable edition of the minutes, papers and letters of the society published as K.T. Hoppen (ed.), *Papers of the Dublin Philosophical Society 1683–1709* (2 vols, Dublin, 2008).

[97] 'Gallery of Illustrious Irishmen, No. XIII. Sir Thomas Molyneux, Bart., M.D., F.R.S.', *Dublin University Magazine*, 18 (1842), 305–27, 470–90, 604–19, 744–64, on pp. 318, 320.

[98] *Correspondence*, vol. 5, pp. 412–13; vol. 6, pp. 253–4.

[99] Allen Mullen, *An Anatomical Account of the Elephant Accidentally Burnt in Dublin ... Together With a Relation of new Anatomical Observations in the Eyes of Animals* (London, 1682); the first work was addressed to Petty.

[100] *Correspondence*, vol. 5, pp. 361–5; vol. 6, pp. 126, 163–6. See also ibid., vol. 5, pp. 296–8, 352.

[101] Royal Society Copy Journal Book, vol. 7, pp. 77–266, passim; R.T. Gunther (ed.), *Early Science in Oxford* (14 vols, Oxford, 1923–45), vol. 10, pp. 91–2, 94, 103, 107, 123, 132, 135, 174.

involving injecting mercury into its veins, which was published posthumously in *Philosophical Transactions* after Mullen's early death in 1691.[102] This must have reminded Boyle of the injection experiments he had been involved in at Oxford in the late 1650s and early 1660s, and, earlier still – taking us full circle – of the anatomical experiments he had carried out with Petty in Dublin in 1654.[103]

By this time, Boyle had more or less retired from the Royal Society, and hence the visits to his London home made by men like Mullen, Molyneux and Wetenhall were rather separate from their visits to the Royal Society. Yet the two clearly mutually reinforced the sense of a community of natural philosophical endeavour that existed. Indeed, the support of Boyle gave a kind of imprimatur to the infant society in Dublin. It is not surprising that when Robert Huntington told Boyle about its inauguration, he added that 'they would be extreamly proud of your directions & Encouragement', and it is interesting to find the society pursuing a Boylean agenda in some of its early investigations, for instance into the nature of freezing or into mineral waters (the second edition of Boyle's book on cold had appeared in 1683, and his treatise on mineral waters in 1685).[104] Boyle also presented copies of his books to its members, as with the copy of his seminal *Free Inquiry into the Vulgarly Received Notion of Nature* that he gave to Robert Huntington in 1686.[105]

But Boyle's links with Huntington, as with Marsh, remind us of a further dimension to Anglo-Irish relations in this period which is easily obscured by a modish emphasis on science. We have seen in Marsh's case how there was also the common bond of their commitment to scholarship, and the intense religiosity which inspired them to collaboration on the Irish Bible project, and these were equally important facets of the intellectual activity of the 1680s. Members of the Dublin society may, indeed, have pursued exegetical and patristic scholarship with as much enthusiasm as science, to judge from the letters between another early member, St George Ashe, and the Dublin-educated scholar Henry Dodwell at Oxford. For these suggest that their concern was not just with 'clearing naturall Theology, after a plainer & more demonstrative method then is yet extant', but also with 'the study of the fathers & controversiall divinity' and with work on 'the Apostolick writers' associated with the edition of the writings

---

[102] Allen Moulin, 'An Account of an Experiment of the Injection of Mercury into the Blood', *Phil. Trans.*, 16 (1686–92), 486–8.

[103] For the Oxford experiments, see R.G. Frank, Jr, *Harvey and the Oxford Physiologists* (Berkeley, CA and Los Angeles, CA, 1980), pp. 169ff.

[104] *Correspondence*, vol. 6, pp. 2–3. For Boylean themes in the society's scientific work, see Hoppen, *Common Scientist*, esp. pp. 102–4, 106, 110 and ch. 5, passim; see also Hoppen, *Papers*, esp. vol. 1, pp. 14–15, 29–33, 35, 51, 124–6, 212–16.

[105] *Correspondence*, vol. 6, p. 198.

of the Fathers planned in the years around 1690 by Dodwell in conjunction with the great Oxford biblical scholar John Mill.[106]

Indeed, in many respects, the intellectual milieu to which Marsh and his colleagues belonged was focused less on London than on Oxford, with whose Philosophical Society, as much as the Royal Society, the Dublin Society was in regular contact.[107] This was a milieu in which Boyle himself was quite at home, where clerics moved easily between learned and scientific activities and between both and the evangelical efforts which were so close to their and to Boyle's heart. Boyle would undoubtedly have approved of the foundation of Marsh's library 10 years after his death: indeed, it is worth noting that, as an alternative to the famous series of apologetic lectures Boyle founded in his will, he apparently considered the possibility of endowing a 'Collection of Bookes tending to the Truth of the Christian Religion'.[108] Again and again, the preoccupations of Boyle and Marsh mutually illuminate one another, and this surely provides a crucial part of the setting for Marsh's initiative in founding the library that bears his name.

---

[106] Bodleian MS Eng. Lett. c. 29, fols. 2, 6b, 7–8, 11–12. See also Adam Fox, *John Mill and Richard Bentley* (Oxford, 1954), pp. 140–41, and Hoppen, *Common Scientist*, p. 88. The controversies referred to were with the Roman Catholics.

[107] See Hoppen, *Common Scientist*, esp. pp. 89–90; Hoppen, *Papers*, vol. 2, passim; Gunther, *Early Science in Oxford*, vol. 4, pp. 29–30 and passim, and vol. 12, pp. 128ff.

[108] *Boyle by Himself and His Friends*, p. xxv.

Chapter 5
The Disquieted Mind in Casuistry and Natural Philosophy: Boyle and Thomas Barlow[1]

In this chapter, I want to address Boyle's interest in casuistry, focusing particularly on a group of casuistical treatises specially written for him in the early 1680s by the divine Thomas Barlow. Through an examination of the liaison between the two men, I hope to explore certain broader issues concerning the intellectual and mental functions that casuistry served for Boyle – what he may have got out of it, how much help a casuist like Barlow was able to give him, and the extent to which such casuistical advice influenced Boyle's scientific and philosophical activity, as against serving a purely pastoral function in relation to his emotional and spiritual life. I hope that this case study may stimulate broader reflection on the significance of casuistry in the intellectual and cultural history of the period.

First, however, it is necessary to fill in the background in terms of Boyle's longstanding concern with cases of conscience. This formed part of a deep piety on his part which was legendary, stemming from a conversion experience that took place during his adolescence and reinforced by the experiences around 1650 that we have already encountered, including his learning Hebrew and other biblical languages at the behest of James Ussher, Archbishop of Armagh. Thereafter, Boyle took a slightly idiosyncratic line on religious matters: he conformed to the established church in 1660, but showed strong sympathy for figures whom many at the time would have dismissed as 'enthusiasts'.[2] The roots of his casuistical interests are perhaps to be traced to his discovery of Stoicism while travelling in Italy and studying at Geneva during his adolescence in the early 1640s; this seems to have inspired the rather generalised ethical assessment and exhortation in which Boyle indulged in such treatises of the 1640s as his

---

1   Previously published in slightly abbreviated form as 'The Disquieted Mind in Casuistry and Natural Philosophy: Robert Boyle and Thomas Barlow', in Harald Braun and Edward Vallance (eds), *Contexts of Conscience in Early Modern Europe 1500–1700* (Basingstoke: Palgrave Macmillan, 2004), pp. 82–99 and 206–10.

2   See *Boyle by Himself and His Friends*, pp. lx–lxxii, 15–16, 27, and *Scrupulosity and Science*, pp. 51–7.

Aretology.[3] Thereafter, Boyle seems to have moved towards casuistry of a more standard type, focused on specific moral dilemmas and their resolution: he makes a brief allusion to such cases in one of his writings in the 1640s,[4] but it is in the 1650s that his interest in such matters is first fully evidenced.

In 1657, Louis du Moulin, the independent divine and intruded Camden Professor of History at Oxford, published a letter he had written at Boyle's behest on the legitimacy of usury and the views on the subject of the French Protestant scholar Claude du Saumaise and others.[5] Just over a year later, Boyle offered a stipend to a rather different figure, the extruded divine Robert Sanderson, so that he could prepare for publication the *Lectures on the Human Conscience* he had given while Regius Professor of Divinity at Oxford in the 1640s: these were generally regarded as among the classics of the genre, and Boyle thus played a personal part in the burgeoning of an English literature on casuistry in the middle years of the seventeenth century. Thereafter, in the early 1660s, Boyle consulted casuistical advisors over the dilemma he faced when granted the impropriations of former abbey lands as part of the Restoration Irish land settlement, which came at the expense of the pastoral needs of the Irish church. Later still, he consulted both casuists and lawyers as to whether he was obliged to take oaths in connection with public offices he held or was offered, such as the Presidency of the Royal Society in 1680.[6] This was the closest Boyle came to the moral dilemmas that many faced in connection with the political changes of the seventeenth century, particularly the debate over the Engagement oath, which inspired much casuistical activity.[7] (It is not, in fact, known whether Boyle himself took this oath.)

In terms of revealing documentation that survives, the climax of Boyle's casuistical concern came right at the end of his life, in the form of the notes he dictated to an amanuensis on interviews with two eminent bishops, Gilbert Burnet and Edward Stillingfleet, about matters on his conscience. This is an extraordinary survival, probably unique in the vividness with which it illustrates

---

[3]   See the texts in *The Early Essays and Ethics of Robert Boyle*, ed. J.T. Harwood (Carbondale, IL, 1991). However, Harwood perhaps conflates the *Aretology* with Boyle's later casuistical concerns to an undue extent. For Boyle's discovery of Stoicism, see *Between God and Science*, pp. 49–50.

[4]   *Works*, vol. 13, p. 118, quoted in *Between God and Science*, p. 62.

[5]   Louis du Moulin, *Corollarium ad Paraenesim suam* (London, 1657), pp. 247–70. I am grateful to Peter Anstey for bringing this text to my attention.

[6]   *Scrupulosity and Science*, pp. 61–8, 74–6, and chs 3–5, passim. See also above, pp. 54, 88–9.

[7]   See Edward Vallance, 'Oaths, Casuistry and Equivocation: Anglican Responses to the Engagement Controversy', *Historical Journal*, 44 (2001), 59–77, and Conal Condren, *Argument and Authority in Early Modern England: The Presupposition of Oaths and Offices* (Cambridge, 2006), ch. 14 and passim.

casuistry from the consumer's, rather than the purveyor's, point of view.[8] The document records the various topics that were on Boyle's mind in the last months of his life and that he raised with his 'confessors', along with their answers. These range from the impropriations of the former abbey lands already referred to; the oaths Boyle had entered into for charitable provision from his estate, which he was afraid that the Williamite campaigns in Ireland might make him unable to fulfil; his landed transactions and the possibility that he might unintentionally have done down his brother and heir the Earl of Burlington; and not least his religious doubts and his fear that he might have committed the unforgiveable Sin against the Holy Ghost.

No comparable items survive among Boyle's manuscript remains, probably due to censorship by their mid-eighteenth-century custodian, Rev. Henry Miles, who was advised by 'a very judicious friend' that such material was 'not suited to the genius of the present age' and who evidently therefore jettisoned it.[9] Fortunately, however, a much bulkier group of documents relating to Boyle's casuistical interests is extant from nearly a decade earlier. These survive, not among Boyle's own papers, but among those of the casuist who wrote them at his behest, Thomas Barlow, formerly Bodley's Librarian and Provost of Queen's College, Oxford, and by this time Bishop of Lincoln. Barlow had had close links with the great casuists of the earlier generation such as James Ussher, some of whose writings he edited, and Sanderson, whom he succeeded at Lincoln and various of whose manuscripts he preserved, and his own casuistical skills were renowned.[10] In 1692, the year after his death, a volume of *Several Miscellaneous and Weighty Cases of Conscience* by him was published, including such classics as 'Mr Cottington's Case', which considered whether a divorce in an Italian court was valid in England. On this, Barlow drew a contrary conclusion from that of the common law, and one contemporary commentator considered that 'a discussion of any one *Case* of *Conscience*, with such variety of Learning ... is not to be found in the Works of any *Forraign* Casuists'.[11] This volume also contained 'The Case of a Toleration in Matters of Religion' that Barlow had written at Boyle's behest in the aftermath of the Restoration, in which he advocated toleration of a wide range of Protestant sects, though excluding Quakers, Socinians and others; however, Barlow was subsequently to think better of this perhaps rather rash

---

[8]   For an edition of these notes with a commentary, see *Scrupulosity and Science*, pp. 72–92.

[9]   Ibid., p. 77.

[10]   See Ussher's *Chronologia sacra* (Oxford, 1660). For Barlow's citation of a casuistical case of Ussher's, see Queen's College, Oxford (hereafter QC), MS 294, p. 66. For Barlow's role in preserving Sanderson's MSS, see QC MSS 216–18, and William Jacobson (ed.), *The Works of Robert Sanderson* (6 vols, Oxford, 1854), vol. v, passim.

[11]   Sir Peter Pett in *The Genuine Remains of that Learned Prelate Dr Thomas Barlow* (London, 1693), sig. A5v. For the MS of the case, see QC MS 289, item 10 (pp. 343–86).

essay, and it fell from sight until published in the very different circumstances of the early 1690s.[12] In addition, Barlow acted as the go-between in connection with Boyle's offer of a stipend to Sanderson in the late 1650s, and this seems to have stemmed from his activity as a casuistical advisor to Boyle at that point, when both were domiciled in Oxford, though only hints of such a liaison survive prior to the episode I am dealing with here.[13]

The context is provided by a sequence of letters between Barlow and Boyle which begins suddenly in July 1681 and continues until 1688. Previously, hardly any letters between the two men survive, perhaps because they mainly communicated orally: Boyle was based in Oxford until 1668, while Barlow was evidently often in London both before and after his elevation to the see of Lincoln in 1675 (though, since the letters refer to continuing face-to-face contact after that point, one is left slightly puzzled as to why they commence when they do). The letters deal with a variety of matters, not least the relations between the French monarchy and the papacy, and the rising persecution of the Huguenots in France, towards which Barlow's tone was notably more apocalyptic than Boyle's.[14] Subsequently, they discussed the events of James II's reign, while a related episode that had arisen in 1684 concerned the 'popish' appurtenances installed at the parish church of Moulton in Barlow's diocese, the subject of another of the 'Cases' published in his 1692 volume, a copy of which he sent to Boyle.[15] The letters also reveal that in 1688 Barlow sent Boyle a copy of the section of his controversial pamphlet *A Few Plain Reasons why a Protestant of the Church of England should not turn Roman Catholic* (1688) – which Boyle was 'pleasd to mention, and kindly approve' – that was suppressed at the behest of the ecclesiastical censors.[16] In addition, although it is not referred to in the correspondence, in 1685 Barlow had shown Boyle his 'animadversions' on *The Prodigal Return'd Home* (1684), a vindication of himself by the Catholic convert

---

[12]   For a discussion, see *Boyle by Himself and His Friends*, pp. lxxi, 73–4. For the MS, see QC MS 289, item 7 (pp. 263–306).

[13]   See *Correspondence*, vol. 1, pp. 370–71, 400–402. For the stimulus to Boyle's patronage of Sanderson being his consultation of Barlow about a case, see Izaac Walton, *Lives*, ed. George Saintsbury (London, 1927), p. 423.

[14]   See *Correspondence*, vols 5–6, passim, esp. vol. 5, pp. 257–61, 331 For a letter of 1667, see ibid., vol. 3, pp. 306–7.

[15]   Ibid., vol. 6, pp. 49–52, 87–8, 89–91 and 204ff., passim. For Boyle's copy of the Moulton case, see BP 4, fols 132–4; the cover sheet, fol. 135, has an endorsement by Henry Miles matching those on the Barlow–Boyle letters, suggesting that at one point it was stored with them.

[16]   *Correspondence*, vol. 6, p. 257. For Barlow's copy of the suppressed section, see QC MS 215, item 1 (fols 3–13).

E. Lydeott, and an ancillary text, of which Boyle had copies made by one of his amanuenses.[17]

The letters also show that Barlow distributed alms to the poor of his diocese on Boyle's behalf,[18] while, from the outset, they are significantly concerned with his casuistical work for Boyle, hence providing a context for the extant treatises. In a letter to Barlow of 7 July 1681, Boyle, who refers to Barlow as 'my Confessor', mentions a 'Paper' he had given Barlow, evidently the 'Case' which Barlow promised to deal with in his reply, dated 12 July that year.[19] It appears from a further exchange the following year that the case in question was not one of those which survive, since in a letter of 26 August 1682 Barlow referred to this earlier text, wanting to borrow it back because he had not kept a copy of it and needed to refer to it in connection with the further task Boyle had by then set him.[20] Indeed, it was perhaps due to this that, from this point onwards, he started keeping copies of what he wrote for Boyle, hence explaining the cache that survives among his papers. From the same letter, it is apparent that this earlier paper was on the question 'De Voto', 'of vows', perhaps dealing with the kind of matters that were to recur in Boyle's interviews with Burnet and Stillingfleet in 1691, or possibly arising from his refusal of the Royal Society Presidency in December 1680.[21]

The paper which Boyle asked Barlow to write in August 1682 was on a different question, one of the classic issues of casuistry, 'An in dubiis, pars tutior sit eligenda?', 'whether in difficult cases the safer course is to be chosen?' Evidently Boyle asked Barlow 'whether that Rule (soe common amongst the Casuists) were universally true, or had some Fallentiae (as the Canonists call them) some faileinges, and particular Cases, in which it did not hold true?'[22] Barlow thereupon set to work, producing the 111-page manuscript treatise on

---

[17]   BP 7, fols 227–32, comprises Barlow's original of his 'animadversions' on *The Prodigal Return'd Home* (in a scribal hand, with additions by him). Ibid., fols 216–21 is a copy by Boyle's amanuensis, Robin Bacon. Ibid., fols 222–6 comprise a copy by Bacon of parts of a further rebuttal of a Catholic convert's justification of himself, QC MS 266, item 6 (fols 78–89); the passages copied are from fols 83 (fol. 226) and 86v–89 (fols 222–5). The Queen's version is a scribal copy, and the copy in the Boyle Papers is evidently derived from a different version, since it has passages which do not appear in the Queen's MS. Barlow refers to the fact that he left Lydiatt's book and his animadversions to it with Boyle when he left London on 4 August 1685 in an endorsement to MS 266, item 6 (fol. 89v). See also Barlow, *Genuine Remains*, pp. 253–4.

[18]   *Correspondence*, vol. 5, p. 328; vol. 6, pp. 199, 228.

[19]   Ibid., vol. 5, pp. 258–9, 261.

[20]   Ibid., p. 332.

[21]   See *Scrupulosity and Science*, pp. 64–8, 82–3.

[22]   QC MS 294, p. 1. Note that, in all quotations from Barlow's MSS, his extensive and slightly erratic underlining has been ignored.

this topic which survives today as Queen's College, Oxford MS 294, dated 10 April 1683. As he put it when accepting the commission the previous August, '(haveinge never ex professo [expressly], seriously studied that point) I find many perplex'd and difficult cases, which will not be easily explicated soe, as to give cleare and full satisfaction';[23] the length and complexity of the final document bears this out.

Almost immediately after receiving it, however, Boyle seems to have asked Barlow to undertake another comparable exercise; conceivably, the stimulus to this may have been an aside Barlow included towards the end of his initial tract about how he was passing over additional scriptural passages concerning the nature of assurance on the part of the regenerate.[24] In any case, with a letter to Boyle of 2 July 1683, Barlow sent him a further 'longe and tedious scrible', which must comprise the second of the three extant papers addressed to Boyle now at Queen's, MS 275, item 6, entitled 'Analecta de Scrupulis et Conscientia scrupulosa', 'Notes concerning scruples and a scrupulous conscience', dated 29 June 1683. Again, he comments on the extant treatise in the letter, remarking:

> I doe heartily thanke you for settinge me upon this subject; which containes very many things, which if they were distinctly examin'd and particularly explain'd, (by men who had time and abilities enough to doe it) might be of exceedinge great use, to direct the conscience and conversation of any, especially of weaker Christians.

He added that, on reviewing it, he felt 'that many things might have beene said with more brevity and perspicuity', evidently reflecting the fact that (as he averred elsewhere) these treatises were written '(as I gott time from other buisines) by fitts onley, and not with a fix'd and continued consideration'.[25]

The third paper, entitled 'Casus, De Conscientia dubitante', 'For the Honourable Robert Boyle Esquire', and not dated, unfortunately cannot be linked to any extant letters. However, though a discrete and self-contained treatise, it is perhaps to be seen as a sequel to the second, to which it refers and certain points within which it can be seen as elaborating with additional material.[26] It now survives as Queen's MS 285, item 5. Thereafter, correspondence shows that Barlow sent a further pair of treatises to Boyle for his comments, but, though he obviously expected these to interest Boyle, Barlow does not state that they were written specifically for him.[27] The first dealt with his argument that the Sin against the Holy Ghost could no longer be committed: since this was a

---

23   *Correspondence*, vol. 5, p. 332.
24   QC MS 294, p. 105.
25   *Correspondence*, vol. 5, pp. 414–5; vol. 6, p. 13.
26   QC MS 285, item 5, p. 23 and passim. The title is on the cover sheet.
27   *Correspondence*, vol. 6, pp. 12–13, 23–4, 28, 30–33.

topic that had come up in the papers specifically written for Boyle and was to recur in Boyle's 1691 interviews, Barlow was right to think that Boyle would find it interesting, though the paper itself is apparently no longer extant. The second paper considered the issue of the fate of the souls of the dead prior to the day of judgment: this was evidently a paper Barlow had written some years previously for Thomas Tenison, the future Archbishop of Canterbury, addressing questions raised by a correspondent of Tenison's, the Norfolk clergyman John Whitefoot. Barlow's argument was that such souls subsisted in a paradisiacal 'midle state' which he linked with the biblical concept of 'Hades', and he saw this as undercutting 'your Popish Canonization and Invocation of Saints, their purgatory, Indulgences, their Jubilees, and such other gainefull tricks to gett money'. He thus recapitulated his idea in his letters to Boyle as well as sending him the treatise in which it was expounded.[28] Whether Boyle was sufficiently interested in the paper to have his own copy made (as Barlow suggested) is not clear: no copy of it survives in the Boyle Papers, and there is no reference to Boyle's comments on it in the subsequent correspondence.

Turning to the three extant treatises specifically written for Boyle, these represent a remarkable survival, not only because of their link with him, but in their own right. Many manuscript cases of conscience survive from seventeenth-century England, and certain classic cases – notably by such masters of the art as Sanderson or Barlow himself – were published at the time.[29] On the other hand, as has often been noted, these cases are dominated by the concerns that were closest to the hearts of the devout gentry to whom such casuists mainly ministered.[30] Thus many of them deal with marital issues and dilemmas concerning consanguinity and the rights of widows. In addition, one finds cases concerning doctrinal and ecclesiastical issues, including the rights of dissenters and relations with Roman Catholics, or constitutional issues, for instance concerning the relationship of regal and divine powers of pardon and oath-taking, especially in a political context. Among the numerous cases that survive

---

[28] QC MS 266, item 7 (fols 90–101); *Correspondence*, vol. 6, pp. 13, 23, 30–32. The first item in QC MS 204 is a further MS on the relevant article in the creed. John Whitefoot was apparently the elder cleric of this name who appears in Venn, *Alumni Cantabrigienses* (4 vols, Cambridge, 1922–27, vol. 4, pp. 390–91) (d. 1699), rather than the younger (d. 1731). Barlow was evidently mistaken in stating in his letter that the figure in question was a Prebend of Norwich.

[29] See Sanderson, *Works*, esp. vol. 5, pp. 1–136; Barlow, *Several Miscellaneous and Weighty Cases of Conscience* (London, 1692), passim.

[30] For a helpful discussion see Keith Thomas, 'Cases of Conscience in Seventeenth-century England', in John Morrill et al. (eds), *Public Duty and Private Conscience in Seventeenth-century England* (Oxford, 1995), pp. 29–56, on pp. 42ff.

among Barlow's papers, it is topics like these that predominate.[31] Indeed, only three other items are even slightly comparable to the Boyle ones, and these may therefore be itemised here, not least since two of them are bound up with one of the Boyle treatises, possibly because Barlow or his executors saw them as related. One, dating from 1675, was a discursive disquisition on a landed gentleman's duty to devote his surplus income to charitable ends, in the course of which Barlow explored the nature of charitable obligation more generally.[32] The second comprised a treatise addressed to a fellow clergyman, Samuel Dugard, dated 10 June 1673, about the rationale of judgments as to the rectitude or 'obliquity' of human actions, accompanied by recommendations for further reading on the subject, though even this devotes a disproportionate amount of space to the rather predictable matter of the Old Testament mandate concerning the legitimacy of the marriage of cousins germane.[33] The third such item, which survives in another volume, is an answer to the question 'How we may know the sinfullnes or innocence of our thoughts', written for Seymour Bowman, a Wiltshire lawyer and MP who was evidently friendly with Barlow at Oxford; in it, Barlow offered general prescriptions on the subject, stressing that the degree of sin depended on the degree of concurrence of will on the part of the individual affected.[34]

Hence Barlow was apparently being quite honest when he told Boyle that the issues that he raised were not familiar ones, and it is worth pausing to emphasize just how unusual such attention to them was. For all three of the papers addressed to Boyle were about what might be regarded as the most difficult parts of casuistry: the issue of 'tutiorism', of reaching the safest decision on a question, and the question of scruples – what kind of doubts these were, and how they should be dealt with. I will return to the significance of this in relation to Boyle in a moment, but here I should stress the rarity of self-contained discussions of such topics. Obviously, matters like these tended to be addressed along with other relevant issues in casuistical treatises, but I am not aware of other separate, sustained discussions of them.

What is more, Boyle's questions inspired Barlow to a vast amount of activity. The paper on tutiorism occupies 111 pages in Barlow's crabbed handwriting, the

---

[31]   See H.O. Coxe, *Catalogus codicum MSS, qui in Collegiis Aulisque Oxoniensibus Hodie Adservantur* (2 vols, Oxford, 1852), vol. 1: 'Catalogus Codicum MSS Collegii Reginensis', pp. 43ff., passim.

[32]   QC MS 275, pp. 1–17.

[33]   QC MS 275, item 5 (pp. 95–125): this was a topic on which Dugard himself had written a book. For Dugard's response dated 27 August 1673, see MS 290, item 9 (fols 39–40).

[34]   QC MS 289, item 4 (pp. 157–200). See also MS 289, item 8 (pp. 307–26) for a discussion of whether madmen were under the moral law. For Bowman, see B.D. Henning, *The House of Commons, 1660–90* (3 vols, London, 1983), vol. 1, pp. 695–6.

other two a further 40 and 41 respectively: in other words, together, they would make up a monograph. Moreover, Barlow attacked his task with gusto and with huge erudition, filling the margins with extensive references to Catholic casuists like Suárez, Filliucius, Escobar or Laymann; to learned debates over the authenticity of different readings of passages in the Bible; and to discussions of the rationale of divine and natural law more generally. They thus exemplify the erudition reflected in the suggestions made by Barlow in his Ἀυτοχεδιάσματα, or, Directions to a young Divine for his Study of Divinity, and choice of Books, &c., published as part of his *Genuine Remains* in 1693.[35]

They also illustrate the nature of Barlow's casuistical method, showing the importance to it of logical analysis, and perhaps making it less surprising that Sanderson, the premier English casuist of the age, had first made his name with a textbook on logic. This is particularly clear in the first treatise, on 'tutiorism', more than half of which was spent on carefully parsing each of the words in the proposition, drawing out their exact meaning, and the alternatives they presupposed. Thus Barlow explained how 'pars' implied alternatives, 'tutior', 'safer', entailed comparison between them, while 'eligenda' implied that there was a significant element of choice. There are also a host of subsidiary definitions, as regarding differing degrees of certainty, or of goodness, while at intervals there are lengthy logical demonstrations, for instance of what to do if two parts of the moral law conflicted.[36]

This treatise also lays out clearly the criteria according to which Barlow believed that decisions on matters of conscience should be made, namely the law of nature and scripture. Like other casuists, Barlow took it for granted that many moral obligations were established by principles of natural law – 'the onely Common-law to all mankind', in his words – which long pre-existed the Christian revelation and could thus be elucidated by recourse to thinkers like Aristotle.[37] But this was critically supplemented by the authority that resided in the scripture; 'the old and New Testament', as he put it:

> doe containe (what we now seeke for) a perfect Rule, to direct all Christians (Clergy and Laity) in all doubts and Cases of Conscience, concerninge good and evil, lawfull or unlawfull, and soe what is to be chosen or refus'd. And it is such a Rule, as all men may, and ought to use, for the regulation of their Actions.[38]

---

[35] Barlow, *Genuine Remains*, pp. 1–121. For references specifically on casuistry, see ibid., pp. 45ff. A separate edition of this work was published in 1699.

[36] QC MS 294, pp. 3ff., 9ff., 50ff., 80ff., 94ff. and passim.

[37] Ibid., p. 39.

[38] Ibid., p. 27.

For the rest, this lengthy treatise comprised a discursive pursuit of the argument through a succession of case studies. Here, as elsewhere, Barlow never missed an opportunity to pick an argument with Roman Catholic writers, not least for what he saw as the discrepancy between their views and the rule of scripture. Indeed, this deep aversion to popery formed one of the mainsprings of Barlow's thought on any subject, and the attack on Roman Catholicism is arguably the predominant theme in his writings as a whole. He also discussed certain matters in which he must have perceived Boyle as having an interest, whether or not he thought he would agree with him, for instance concerning the right of Protestant dissenters to appeal to their conscience in opposition to the rulings of the magistrate (Barlow's sympathies were with the magistrate).[39] And he recounted a series of semi-fictional cases to illustrate the principles he was expounding, for instance, that of a London gentleman who, mindful of the 'Divine precept' that his servants and livestock as well as he should rest on the sabbath, wondered whether it was 'pars tutior', 'the safer course', to go to church in a sedan chair or by coach, the former involving two men, the latter only one man, but also two horses: 'I answer, that in this Case (if his doubt continue) there is a safer way, then either of the two proposed: lett him goe on foote, and take his servants with him; and then he may be sure he sins not.'[40]

The opening section of the second treatise, 'about the name and nature of Scruples',[41] again virtually constitutes a general treatise on conscience, since Barlow clearly saw the need to place the issue of the legitimacy of scruples in the context of how decisions in matters of conscience should properly be made, and hence how such doubts might relate to them. Echoing the commonplace views of early modern casuists, he thus differentiated five states of the conscience: 'recta', 'erronea', 'probabilis', 'dubia' and 'scrupulosa'.[42] Interestingly, he had no time for the view that the 'probable' conscience was a distinct category, the famous crux which Pascal stressed in his notorious critique of Jesuit casuistry in the *Provincial Letters* (1656–57). Barlow comparably dismissed the status of scruples, of which he helpfully provided an etymology, explaining how the word derived from a little stone caught in a shoe, and the inconvenience and mild pain that derived from that. These were 'little doubts and vane feares, such as have very little or noe reason to induce them', which arose 'consequent to the determination and Judgement of conscience', in contrast to truly founded

---

[39]   Ibid., pp. 20–21, 70–72.

[40]   Ibid., pp. 65–6.

[41]   QC MS 275, p. 135 (item 6, p. 1)

[42]   Ibid., p. 137 (item 6, p. 3). For parallels, see, for example, Jeremy Taylor, *Ductor Dubitantium* (London, 1660), book 1, or the summary of Balduin's views in H.-D. Kittsteiner, 'Kant and Casuistry', in Edmund Leites (ed.), *Conscience and Casuistry in Early Modern Europe* (Cambridge, 1988), pp. 185–213, on p. 199n.

doubts which might have prevented a decision being reached in the first place.[43] Hence the proper way to deal with them was to dismiss them, since to act in accordance with them meant following one's fears and surmises, rather than the dictates of moral prudence. Then, citing Balduin, Barlow proceeded to give seven causes from which they arose, including ignorance, melancholy and temptation by the Devil, and to advocate appropriate remedies, though not all of these were explained in equal detail: indeed, it may have been this that stimulated Boyle to request further elucidation, thus perhaps explaining Barlow's provision of a sequel to this treatise. On the other hand, many associated issues were explored, for instance whether scruples arose following a good or a bad decision: Barlow was inclined to favour the former, on the grounds that the Devil, who inspired them, had no reason to disturb a decision that was morally wrong in any case.[44]

At this point, the treatise rather changes gear, as Barlow gave a lengthy account of a case that had arisen much earlier in his pastoral career, which he clearly saw as germane to Boyle's enquiry. This concerned a young gentlewoman who had come to Barlow for help when he was at Oxford, and it provides an interesting illustration of casuistical activity on his part which would have gone unrecorded had it not been enshrined here. It also illustrates how casuistry overlapped with the broader pastoral concerns of the Anglican clergy, something which excessive stress on cases of conscience dealing with extraordinary dilemmas such as those concerning political oaths or marital contracts tends to obscure: it is thus significant that the planned casuistical treatise which materialised as the *Christian Directory* by the Presbyterian divine, Richard Baxter, was intended by its projectors to be as much a body of practical divinity as a handbook of difficult cases.[45]

Barlow's narrative was as follows. The woman it concerned 'was (as I really beleive) a pious and good Christian', but she 'was exceedingly troubled with scruples of Conscience; soe that she did not, and (as she said) dared not pray'. Barlow and she therefore 'had for divers dayes, severall and longe discourses about her scruples; and I indeavour'd to persuade and convince her, that noe

---

[43] QC MS 275, p. 139 (item 6, p. 5).

[44] Ibid., pp. 145–6, 151–2 and passim (item 6, pp. 11–12, 17–18).

[45] For this project, see J.M. Batten, *John Dury: Advocate of Christian Reunion* (Chicago, IL, 1944), pp. 52–3, 92, 131; John Dury, *An Earnest Plea for Gospel-Communion* (London, 1654), pp. 79ff; Richard Baxter, *A Call to the Unconverted* (London, 1658), sigs A2v–A3; Matthew Sylvester (ed.), *Reliquiae Baxterianae* (London, 1696), i, 122. It is perhaps worth noting that the central figure in it was James Ussher, Boyle's former mentor, who might even have inspired Boyle's casuistical interests in the first place, though there is no direct evidence of this. For Ussher's casuistical activity, see Nicolas Bernard, *The Life and Death of ... James Ussher* (Dublin, 1656), pp. 93–4.

such doubts or scruples as she pretended, should hinder, but rather incourage her to pray'.[46] He explained how (I quote slightly selectively):

> Amongst other things, she was continually examininge her thoughts, words and actions; and she said, her faileings in doeing her duty were soe many and great, that she was affraid to goe to God in prayer ....
>
> Then I told her, that she must pray. For that was the sure and infallible meanes, which our most gratious God had appointed for our deliverance from all such troubles [At this point, he quoted a series of biblical passages in support on this.] This did somewhat, but not altogether satisfy my scrupulous Gentlewoman. For she reply'd; that she was affraid, that if she did pray, God would not heare her for her sins were very many and very great, and the scripture told her, That God heareth not sinners.[47]
>
> In answer to this, I told her. 1°. That it was the malitious cunninge and method of Satan, to worke presumption in some, and dispaire in other sinners ....
>
> 2°. But when grace had wrought true repentance in any body, and a reall sense of sin; then it was the Devills designe and indeavour, to represent his sins, in the most horrid shape and magnitude he could; suggestinge that they had sin'd against the holy-Ghost, and soe beyond all hopes and possibility of pardon; that he might drive them to dispaire and therefore possibly, her sins were neither soe many, nor soe great, as her troubled conscience seem'd to represent them. But it might be that Satan (by his malitious and subtle suggestion) made them seeme more and greater then indeede they were, that soe he might disquiett and hinder her, to use the appointed meanes (prayer) to procure pardon.
>
> 3°. But admitt that her sins were as many and as great as she imagind (onely the sin against the Holy-Ghost excepted, which I told (and satisfy'd) her, could not now be committed by any) yet she had noe reason to be affraid or doubt of pardon, if she heartily pray'd for it.[48]

Laboriously, he sought to convince her of God's assurance of grace to those who sincerely sought repentance, recounting successive meetings in which she came back with fresh objections, to each of which he recorded his answers and arguments at length, until finally 'the (before) scrupulous Gentlewoman, seem'd fully satisfyed; and by the mercy of God, (from whom all blessings come) all her scruples vanished, and her conscience (to her great comfort) was quietted'.[49]

---

[46]   QC MS 275, p. 155 (item 6, p. 21).

[47]   Joh. 9. 31 [Barlow's note].

[48]   Ibid., pp. 155–7 (item 6, pp. 21–3): in Barlow's original, 'be' is actually written 'ber' under point 2.

[49]   Ibid., p. 174 (item 6, p. 40).

The third of Barlow's treatises for Boyle was, as already noted, something of a sequel to the second, dealing with the issue of religious doubt by retailing a further relevant instance, in this case of a clergyman in his diocese who was worried that prayer was inefficacious in a man of wavering faith. In recounting the case for Boyle, Barlow explained how:

> because the Case concern'd doubtinges and scruples of Conscience (a subject of which you and I have often discours'd) I have taken the confidence, to give you the summe, and a short (yet a distinct) account of it, and desire your Opinion, and Animadversions upon those particulars, to which my friend and I, have assented, and (at present) receaved as truthes. For I know, that your Judgement is such, that if there be any mistakes, (as I feare there may) that you will soone see them, and then your charity is such, that you candidly discourse to us, (who will thankfully acknowledge that favor) what, and where those mistakes are.[50]

It is perhaps worth noting here that, even in his first treatise, Barlow had ended with the 'case' of Titius, in other words Boyle himself, and whether he 'should not admonish me of such things as he believes erroneous'.[51] As certain passages in the letters between Boyle and Barlow suggest, this was not an entirely one-sided encounter, with Barlow as Boyle's mentor, but involved an element of discussion: indeed, in one Barlow himself described Boyle as one 'whom I dare make my confessor'.[52]

As in the second treatise, Barlow was here provided with an opportunity to drive home his evangelical message. He told Boyle how he had explained to his anonymous confident – 'a learned Divine, and (as I have great reason to beleive) a sincere penitent, and a pious person' – that doubt was commonplace among even the most prominent Christians, due to 'Satan's great power and policy, the allurements of the wicked world, and their owne corruptions (the unhappy remaines of the old man, as scripture calls them) to struggle with'.[53] Even figures as eminent as the apostles or 'the best of Gods saints, doe not obtaine such a measure of faith, but are sometimes troubled with doubts and feares, which disquiet their mindes, and (at least) lessen their peace of Conscience'.[54] But set against this was God's dispensation and Christ's atonement, which Barlow reiterated again and again, explaining how: 'notwithstandinge those feares and doubtings, which may be, and sometimes are, even in the best men; yet the Promises we have spoken of, are (in themselves) a most firme and Infallible

---

[50]   QC MS 285, item 5, p. 1.
[51]   QC MS 294, p. 110.
[52]   See *Correspondence*, vol. 5, p. 415; vol. 6, p. 257. Cf. vol. 6, p. 30.
[53]   QC MS 285, item 5, pp. 1–2.
[54]   Ibid., p. 34.

ground of assurance and confidence, and abundantly able to exclude all feares and doubtings'.[55] As before, all this was interspersed by learned disquisitions on the meaning of biblical passages, and by careful distinctions – for instance, between God's absolute and conditional promises, or between doubts that were and those that were not consistent with true faith.[56] Towards the end, however, Barlow's text rises to a climax, almost like a sermon, in which the message of Christ's atonement is rammed home.

These texts are interesting in themselves for what they reveal about Barlow and his pastoral activity in the broader context of late seventeenth-century England. Yet they are also interesting in relation to Boyle – if obliquely, since we need to judge how directly we can extrapolate from the issues Barlow dealt with to Boyle's personal and intellectual life. (In this regard, they differ from the 1691 interviews with Burnet and Stillingfleet, which were concerned directly with Boyle's spiritual and moral state, since, as disquisitions on related issues, the link between these and Boyle's own experiences is only implicit.) However, it is clear that religious doubt was something Boyle experienced throughout his life. He wrote frankly about this in his early autobiographical treatise *An Account of Philaretus during his Minority*, reiterating in a slightly later work how 'He whose Fayth hath never had any Doubts, hath some cause to Doubt whether he hath ever had any Fayth.'[57] At the other end of his life, the reality of Boyle's religious doubts comes across strongly from the interviews with Burnet and Stillingfleet, and these also show that Barlow's assurances to him of the impossibility of committing the Sin against the Holy Ghost had not quietened his anxieties on that score.[58] It must have been at least partly because he knew that this was a matter of particular concern to Boyle that Barlow thought it appropriate to include the lengthy narrative of the Oxford gentlewoman's struggle with such doubts in his second treatise.

On the other hand, Barlow himself disclosed that he, too, was subject to doubt: indeed, at times in his third treatise he implied that such doubts were almost a sign of true faith, though elsewhere he backed away from this and implied that the greatest Christians transcended such tendencies.[59] Be that as it may, we clearly see here something of the natural history of religious experience in seventeenth-century England: contrary to those who have conflated such revelations with the open irreligious opinions of sceptics, doubt was clearly integral to the religious experience of the devout, and was presumably normally a

55   Ibid., p. 21.
56   Ibid., pp. 15ff., 35 (marginal note) and passim.
57   *Boyle by Himself and His Friends*, pp. 15ff.; *Works*, vol. 13, p. 181.
58   *Scrupulosity and Science*, pp. 83, 88–9, 90–91.
59   MS 285, item 5, pp. 2, 22–3, 31–2, 34. Cf. MS 275, pp. 141–3 (item 6, pp. 7–9).

private matter.[60] Indeed, one has a slightly uncomfortable sense of eavesdropping on confidential matters in documents like these, and, partly for this reason, it is hard to judge whether Boyle was abnormal in the degree to which he was the victim of feelings of this kind – even if they may have made him a doughtier defender of religious truth against its perceived antagonists than might otherwise have been the case, for instance in his discussion of miracles.[61]

In any case, if one looks at the interviews as a whole, including the lengthy discussion of 'tutiorism' as well as that of 'scrupulosity', what is striking is their suggestion that, for Boyle, the problem was not so much of outright doubt as of indecisiveness. Boyle seems to present an extreme example of the 'perplexing uncertainty' to which Conal Condren has seen early modern casuistry as being prone more generally.[62] In Boyle's case, it is clear from his request for advice from Barlow that he wanted guidance on the criteria by which he could achieve a degree of conviction on such knotty matters. This need not have been absolute certainty, of the kind associated with mathematics, but merely a moral certainty of the kind Boyle considered perfectly acceptable on issues of this kind.[63] But this was clearly a goal which was obstructed by the recurrence of second thoughts and complications of the kind that 'scruples' exemplified.

Indeed, here one recognises the distinctive figure of Boyle the natural philosopher, the man whose 'penchant for equivocation' is notorious – often meaning that the final versions of his books are less clear than the preliminary ones from which they derived – and whose proneness endlessly to qualify his statements contributed to his 'inability to bring a sentence to a successful conclusion' (as one author has put it).[64] Linked to this is Boyle's well-known

---

[60]   For such conflation, see, for example, Keith Thomas, *Religion and the Decline of Magic* (London, 1971), pp. 166ff. For the distinction made here, see Michael Hunter, *Science and the Shape of Orthodoxy* (Woodbridge, 1995), pp. 329–30.

[61]   See *Scrupulosity and Science*, p. 83. See also R.S. Westfall, *Science and Religion in Seventeenth-century England* (New Haven, CT, 1958; reprinted Ann Arbor, MI, 1973), pp. 87–92; Rosalie Colie, 'Spinoza in England, 1665–1730', *Proceedings of the American Philosophical Society*, 107 (1963), 183–219; R.M. Burns, *The Great Debate on Miracles: From Joseph Glanvill to David Hume* (Lewisburg, PA, 1981), pp. 51–7; Peter Dear, 'Miracles, Experiments and the Ordinary Course of Nature', *Isis*, 81 (1990), 663–83; J.J. MacIntosh, 'Locke and Boyle on Miracles and God's Existence', in Michael Hunter (ed.), *Robert Boyle Reconsidered* (Cambridge, 1994), pp. 193–214; J.J. MacIntosh (ed.), *Boyle on Atheism* (Toronto, 2005), esp. pp. 261ff.

[62]   Condren, *Argument and Authority*, p. 174, and see ch. 8, passim.

[63]   See, for example, *Works*, vol. 8, pp. 281–2. For background, see H.G. van Leeuwen, *The Problem of Certainty in English Thought 1630–90* (2nd edn, The Hague, 1970), pp. 91ff.

[64]   Lawrence M. Principe, *The Aspiring Adept: Robert Boyle and his Alchemical Quest* (Princeton, NJ, 1998), p. 28; Michael Hunter (ed.), *Robert Boyle Reconsidered* (Cambridge, 1994), p. 12; R.S. Westfall, quoted in *Boyle by Himself and his Friends*, p. lxxv.

indecisiveness on almost any major issue that confronted him, whether in theology or natural philosophy, in contrast to bolder figures who were quicker to draw conclusions. This was perhaps put most clearly in Boyle's well-known exposition of his natural philosophical method in the 'Proemial Essay' to his *Certain Physiological Essays* of 1661, addressed to his nephew Richard Jones, in the guise of 'Pyrophilus'. In the context of the emphasis that he there placed on open-mindedness and the avoidance of systems, Boyle explained:

> Perhaps you will wonder, *Pyrophilus*, that in almost every one of the following Essays I should speak so doubtingly, and use so often, *Perhaps, It seems, 'Tis not improbable*, and such other expressions to argue a diffidence of the truth of the Opinions I incline to, and that I should be so shy of laying down Principles, and sometimes of so much as venturing at Explications. But I must freely confess to you, *Pyrophilus*, that having met with many things of which I could give my self no one probable cause, and some things of which several Causes may be assign'd so differing, as not to agree in any thing unless in their being all of them probable enough, I have often found such Difficulties in searching into the Causes and Manner of things: and I am so sensible of my own Disability to surmount those Difficulties, that I dare speak confidently and positively of very few things, except of Matters of fact. And when I venture to deliver any thing by way of Opinion, I should, if it were not for meer shame, speak yet more diffidently than I have been wont to do.[65]

Of course, up to a point, the issue could be avoided by recourse to 'matters of fact', as Boyle implies: indeed, whereas this strategem on Boyle's part has famously been seen by Steven Shapin and Simon Schaffer as his response to external controversies, it might be more fruitful to see it as a way of resolving the tensions in his own mind.[66] But, for all the value of matters of fact, they still had to be assessed and evaluated, and the respite they provided from debate was only temporary. Though sometimes Boyle might 'content my self at present to have faithfully delivered the *Historical* part of these *Apparences*, without making, at least at this time, any Reflections on them', in the knowledge that 'these Experiments will occasion among the *Virtuosi* several *Queries* and *Conjectures,* according to the differing *Hypotheses* and Inquisitions, to which men are inclined', he could not continue this indefinitely.[67] Instead, he had to return to the difficult business of assessing how his findings might relate to the philosophical schools of his day,

---

[65]　*Works*, vol. 2, p. 19.

[66]　Steven Shapin and Simon Schaffer, *Leviathan and the Air-pump: Hobbes, Boyle and the Experimental Life* (Princeton, NJ, 1985), esp. ch. 2.

[67]　'New Experiments concerning the Relations between Light and Air' (published in *Philosophical Transactions* in 1668), *Works*, vol. 6, p. 14.

where his characteristic indecisiveness reasserted itself, becoming all the more marked in highly controversial areas. Thus, we have his convoluted attempt to adjudicate between the rival claims of Galenic and Paracelsian physicians in *The Usefulness of Natural Philosophy* (1663), while, though he later strengthened his resolve in this field sufficiently to plan and at least partially write an open attack on contemporary medical practice, pusillanimity then again prevailed, leading him to suppress the work, largely because of his doubts as to how far he could be sure that his criticisms were more valid than the therapy against which he pitted them.[68] Later, to take a single further example, his attempts to decide between Cartesian and other views on the vexed question of final causes in natural philosophy led to similar convolution.[69] Indeed, it is worth noting that even Boyle's continued proclivity for the dialogue form in his later writings may partially reflect the fact that this enabled him to express a variety of points of view without necessarily resolving them.

This, however, merely states a parallelism between Boyle's proclivities in natural philosophy and in matters of conscience without necessarily postulating a link between the two. But can one find a causal connection between his activities in the two spheres? Certainly, as I argued some years ago, Boyle's profuse experimentation can itself be seen as a transference of his scrutiny of his conscience to the laboratory, involving the examination of a matter again and again until he was satisfied.[70] He told Bishop Burnet that 'He made Conscience of great exactnes in Experiments', and he repeatedly uses the language of casuistry in describing the way in which he was impelled from one experiment to another, referring in an experimental context to his 'scruples' and to the 'cases' that preoccupied him.[71] Yet, as he explained elsewhere, the recourse to further data did not necessarily resolve an issue as against further complicating it, since:

> an Inquisitive Naturalist finds his work to increase daily upon his hands, and the event of his past Toils, whether it be good or bad, does but engage him into new ones, either to free himself from his scruples, or improve his successes. So that, though the pleasure of making Physical Discoveries, is, in it self consider'd, very great; yet this does not a little impair it, that the same attempts which afford that delight, do so frequently beget both anxious Doubts, and a disquieting Curiosity.[72]

---

[68]   *Scrupulosity and Science*, ch. 8, and see above, pp. 4, 13.

[69]   See Timothy Shanahan, 'Teleological reasoning in Boyle's *Disquisition about Final Causes*', in Hunter (ed.), *Robert Boyle Reconsidered*, pp. 177–92.

[70]   See *Scrupulosity and Science*, ch. 3.

[71]   *Boyle by Himself and his Friends*, p. 28; *Scrupulosity and Science*, p. 69; *Works*, vol. 4, p. 212.

[72]   Ibid., vol. 8, p. 58.

Is there any evidence that, through casuistry, Boyle was assisted in resolving these 'anxious Doubts' in natural philosophy, as it was obviously the intention that they should be stilled in his religious life? In particular – in the context of the documents I have considered in this chapter – is there any evidence that Boyle might have obtained such help from Barlow? At the outset, I should say that there is no reason why this should necessarily be the case. We are arguably unduly prone to presume that people, whether modern or early modern, are all joined up, and that each part of their life must be inextricably linked to each other. In fact, however, Boyle himself informs us that there were aspects of his religious commitment which he consciously refrained from writing about in his published books, ignorance of which has led to a serious misrepresentation of his religious position by the otherwise admirable scholar R.S. Westfall.[73] There is no reason why Boyle should not have looked to Barlow for help on religious matters in their own right, especially the kind of essentially private matters on which I have already commented.

It is also worth noting that Boyle was unlikely to get much help from Barlow on the key issues to do with the interpretation of nature, or the relationship of natural philosophy and religion, that so preoccupied him. Barlow's literary remains reveal an intellectual outlook so different from Boyle's that it is in some respects surprising that the two men got on as well as they did. As already noted, Barlow was preoccupied by doctrinal and ecclesiastical issues, particularly those dividing the Protestant and Roman Catholic churches. Yet, though Boyle had copies made of some of Barlow's anti-Catholic writings, which he preserved among his papers, his own writings almost entirely lack the frenetic anti-Catholic tone of Barlow's. Even within the Protestant–Catholic debate, there were issues which seem not to have interested Barlow although they might have struck a chord with Boyle, for instance the continuing role of miracles, a theme in Lydeott's *The Prodigal Return'd Home* which Barlow entirely ignored in his 'animadversions' on the work in favour of issues concerning tradition and ecclesiastical authority.[74] More broadly still, one might note the difference between the churchmanship of the two men, Barlow being 'a thorough paced Calvinist', in the words of the antiquary, Anthony Wood, in contrast to Boyle's more liberal and eclectic outlook.[75]

---

[73] See *Scrupulosity and Science*, pp. 239–40, commenting on Westfall, *Science and Religion in Seventeenth-century England* (New Haven, CT, 1958), pp. 125–6.

[74] E. Lydeott, *The Prodigal Return'd Home* (n.p., 1684), esp. pp. 308–66. For Barlow's 'animadversions', see above, n. 17. For Boyle's interest in miracles see above, n. 61.

[75] Anthony Wood, *Athenae Oxenienses*, ed. Philip Bliss (4 vols, London, 1813–20), vol. 4, p. 335. On Boyle's and Barlow's churchmanship, see *Boyle by Himself and His Friends*, pp. lxx–lxxii. On Boyle's religious outlook, see also Peter Anstey, 'The Christian Virtuoso and the Reformers: are there Reformation Roots to Boyle's Natural Philosophy?', *Lucas*, 27–8 (2000), 5–40.

Intellectually, Barlow was a dyed-in-the-wool scholastic: indeed, scholasticism was central to his casuistical method, and the way concepts and words were parsed owed much to scholastic terminology and logic. In his suggestions for 'a Library for Younger Schollers', Barlow's 'take' on natural philosophy was almost entirely scholastic, although he made a nodding allusion to the new philosophy; though this was admittedly a much earlier work, a similar outlook is indicated by the passing allusions to such matters in the treatises written for Boyle, as when he saw degrees of faith as analogous to degrees of qualities.[76] On one of the few occasions when he discussed natural phenomena in a letter to Boyle, concerning the comet of 1682, Barlow showed himself very traditional in his expectation of what 'art' could achieve in terms of the understanding of comets, limiting this to their motion, magnitude and altitude. He was adamant that the issue of 'how they come to appeare, and of what matter they are made', as also 'how, and by what meanes, they disapeare', were as 'impossible to be knowne, by any Art' as 'Their *signifycation*, what they (in particular) portend' – a monitory significance which he considered as axiomatic in this case as in that of earthquakes two years later.[77] It is also worth noting that Barlow was probably the author of a swingeing attack on William Petty's *Discourse concerning Duplicate Proportion* (1674), partly on the grounds that it was logically inept, and partly that its mechanistic viewpoint was potentially atheistic, going on to see the new science as a whole as a 'popish' plot, liable to draw its students away from the anti-Catholic controversies that bulked so large in his own intellectual agenda.[78]

Boyle gave Barlow copies of several of his books, but those on natural philosophy which bear Barlow's ownership inscription show little sign of having

---

[76] [Thomas Barlow], '*A Library for Younger Schollers*.' *Compiled by an English Scholar-Priest about 1655*, ed. Alma de Jordy and H.F. Fletcher, Illinois Studies in Language and Literature, vol. 48 (Urbana, IL, 1961), pp. 3–4; QC MS 285, item 5, p. 23.

[77] *Correspondence*, vol. 5, pp. 331, 430.

[78] Barlow, *Genuine Remains*, pp. 151–6, 157–9; for an MS version of the first, see Royal Society Early Letters P.1.32. For a discussion of its authorship, see L.G. Sharp, 'Sir William Petty and Some Aspects of Seventeenth-century Natural Philosophy' (unpublished Oxford D.Phil. thesis, 1977), pp. 318–19, 412. However, I am not convinced by Sharp's arguments against Barlow's authorship. In fact, contrary to Sharp's statement, Barlow was close to Anglesey (see, for example, *Genuine Remains*, pp. 190–201, 302–8, 309–11); and although he was sent a copy of the book himself, it is quite plausible that it was because he was solicited to do so by Anglesey that he expressed his view on it; it is also plausible that he sought to avoid personal offence by using the pseudonym 'L.M.' in the printed version. In addition, as noted in Michael Hunter, *Science and Society in Restoration England* (Cambridge, 1981), p. 138n., the second letter, which refers back to the first, is definitely by Barlow. It should be noted that Sharp has no real alternative suggestion as to who wrote the disputed letter. For further discussion, see Rhodri Lewis (ed.), *William Petty on the Order of Nature: An Unpublished Manuscript Treatise* (Tempe, Arizona, 2012), p. 4 and n. 11, who comes to the same conclusion as me (and identifies 'J.B.' as Barlow's friend, Sir John Berkenhead).

been read, even if ones on philosophical and theological topics seem to have interested him more.[79] In particular, Barlow made extensive pencil markings in his copy of Boyle's *Of the High Veneration Man's Intellect Owes to God* (1684–85), a work combining pious reflection on God's attributes with data about His power drawn from the new science: this was evidently written or revised within a few years of its publication, and hence provides an interesting setting for Boyle's casuistical encounter with Barlow in these years. Barlow was so impressed by the astronomical data included in the work that he wrote out a page of slightly naïve notes on them, as if for use to impress his congregation in the manner of a modish modern cleric. These 'Observations (which to many seeme incredible) grounded upon the Authority and reasons of the best Mathematicians (who onely are competent Judges of such things)' included the fact that the earth was 26,000 miles in circumference, or that the Sun was 160 times larger than the Earth, if not ('according to the Calculation of some later, and more exact Artists') 8,000 or 10,000 times.[80] On the other hand, although these notes illustrate even Barlow being momentarily caught up in the excitement of the new philosophy, this is arguably the exception that proves the rule, his commonsensical but old-fashioned approach to such matters being perhaps more typically revealed by an example he gave to illustrate relative degrees of certainty in his first casuistical treatise for Boyle:

> Travellinge one time in Yorkshire, (pray pardon the extravagance) I overtooke an honest Miller; amongst other things, we came to talke of the sunne. I asked the Miller how big he thought the sunne was? He reply'd, (which was true) that he knew not. I ask'd againe, if he thought it was as big as that ground we were in? (which was not any way 100 paces over) he answered that was much; yet (for ought he knew) it might be as big as the ground we were in. I demanded againe, whether he thought the sunne was as big as Yorkshire? to that he quickly reply'd; Noe (by my faith) it could not be soe big, for that was the greatest shire in England.[81]

Even so, could casuistry – either specifically through Barlow, or more generally – have helped Boyle in some of the conceptual issues involved in natural philosophy? In particular, there is the issue of the source of such facets of Boyle's philosophy of nature as his consideration of different levels of certainty, or his

---

[79]   See, for example, Bodleian Library, Oxford, A.20.9 Linc (*Defence* and *Examen*, 1662) or 8° C 352 Linc (*Icy Noctiluca*, 1682), which have no markings. 8° C 336 Linc (*Specific Medicines*, 1685) has some at the start, but they gradually peter out and there are none after p. 17. 8° B 175 Linc (*Things above Reason*, 1681) is quite heavily marked: for Barlow's misapprehension about the authorship of this book, see *Works*, vol. 9, p. xxxi.

[80]   8° C 426 Linc.

[81]   QC MS 294, p. 6.

criteria for accepting or rejecting proof.[82] It is true that these were issues that were widely discussed in the literature of casuistry, and disquisitions on related points occur in the treatises Barlow wrote for Boyle. Thus we have seen how, particularly in connection with his consideration of tutiorism, Barlow devoted much space to assessing differing types of certainty; indeed, it is perhaps worth noting that it now seems likely that Boyle used the concept of 'probability' in the traditional, casuistical sense of 'worthy of approbation', rather than the more modern, quantitative sense that was just beginning to emerge in his time.[83] But here one needs to be cautious in assessing the role of casuistry as against the broader philosophical and legal traditions of which casuistry was itself an outgrowth, and it would be rash to postulate a direct casuistical influence on Boyle in such matters when he could have found similar discussions in other authors going back to the time of Aristotle. The same is true with issues of proof, since Boyle often uses explicit legal analogies in this connection, and even the concept of a 'case' straddled both casuistry and the law.[84] In view of our sheer ignorance of Boyle's ownership and reading of books – beyond the evidence from citations in his published works – it would be unwise to overemphasise the likely influence on him of these texts by Barlow just because we have them and know that they were written for Boyle.[85]

Having said this, however, what mental habits is an encounter like this likely to have encouraged or reinforced in Boyle? Undoubtedly, Barlow's treatises – like other cases of conscience, such as those of Sanderson – show great clarity of thought. In each, an issue is approached through a series of numbered sections and subsections, setting out terms of reference, analysing terms, subdividing and – often – offering qualifications. Indeed, it is worth speculating that this very rigour may have been something of a solace to Boyle, in that Barlow's systematic exploration of every contour of an issue of conscience, so that no consideration escaped his casuistical eye, may have provided comfort which any amount of wise counsel by non-casuistical friends could not. Clearly, the practice aided clarity of thought, forcing those engaged in it to lay out their criteria and to work through them clearly and systematically, even if this sometimes degenerated into logic-chopping. Indeed, though the whole rationale of the exercise may

---

[82]   See *Scrupulosity and Science*, pp. 69–70 and works there referred to. See also Rose-Mary Sargent, *The Diffident Naturalist: Robert Boyle and the Philosophy of Experiment* (Chicago, IL, 1995).

[83]   See ibid., pp. 55 and 243–4 n. 82, applying in Boyle's case the critique of Ian Hacking, *The Emergence of Probability* (Cambridge, 1975), in D.L. Patey, *Probability and Literary Form* (Cambridge, 1984), pp. 266ff.

[84]   See, for example, *Works*, vol. 8, p. 285. See also Sargent, *Diffident Naturalist*, ch. 2.

[85]   On Boyle's library, see Iordan Avramov, Michael Hunter and Hideyuki Yoshimoto, *Boyle's Books: The Evidence of his Citations*, Robert Boyle Project Occasional Papers No. 4 (London, 2010).

seem to us rather repellant – and alien to the post-Cartesian stance on issues to do with logic and epistemology espoused by a man like Boyle – it is worth recalling the extent to which an appeal to formal logic as an aid to clarity of thought continued throughout Boyle's period. This was seen, for instance, when his Oxford colleague John Wallis presented his *Institutio logicae* to the Royal Society in 1685, emphasising how logic helped us:

> to manage our reason to the best advantage, with strength of argument and in good order, and to apprehend distinctly the strength or weakness of another's discourse, and discover the fallacies and disorder whereby some other may endeavour to impose upon us, by plausible but empty words, instead of cogent arguments and strength of reason.[86]

An almost trivial example of the way in which Barlow felt that Boyle's mode of presentation could have been improved is suggested by his copy of *High Veneration*, in which he wrote the figures '1' and '2' in the margin at one point in the text where Boyle buried his 'first' and 'second' points within continuous prose, and his copies of other works by Boyle are similarly endorsed.[87]

Equally important was the extent to which it was in the nature of casuistry to focus analysis on specific cases, often differentiating between what might be true *de toto genere*, and what applied in particular circumstances.[88] Boyle must have found resonances in this for the stress on particulars as against generalisations which is central to his entire stance in natural philosophy. Indeed, one is reminded of Albert Jonsen and Stephen Toulmin's view that the ethical issues addressed by casuistry are quintessentially practical rather than theoretical, analogous less to geometry than to clinical medicine (and hence to Boylean experimental philosophy).[89] All this makes it at least plausible that Boyle's mode of thinking was affected by reading material similar to the treatises with which

---

[86]   Quoted in Mordechai Feingold, 'The Humanities', in Nicholas Tyacke (ed.), *The History of the University of Oxford, Volume IV: Seventeenth-century Oxford* (Oxford, 1997), pp. 211–357, on p. 282. For the continuing vitality of traditional logic in this period, see Feingold's discussion in ibid., pp. 276ff., passim.

[87]   See pp. 94–5 of the copy referred to in n. 80 above. Similar markings appear on p. 4 of Barlow's copy of *Specific Medicines* and pp. 5–7 and 34–5 of *Things above Reason*.

[88]   See, for example, Sanderson, *Works*, vol. 5, p. 127. Echoes of such considerations can be found in Boyle's writings, especially such methodological treatises as *The Excellency of Theology, compar'd to Natural Philosophy* (1674) or *The Reconcileableness of Reason and Religion* (1675), as when he differentiated 'Reason consider'd in it self' from 'Reason consider'd in the Exercise of it, by this or that Philosopher': *Works*, vol. 8, p. 255 and passim.

[89]   See A.R. Jonsen and Stephen Toulmin, *The Abuse of Casuistry: A History of Moral Reasoning* (Berkeley, CA and Los Angeles, CA, 1988), esp. ch. 1.

Barlow provided him, even if it has to be admitted that these specific treatises of the 1680s came too late to have a formative influence on him.

In conclusion, it perhaps needs to be stressed that it was in the nature of casuistry – Reformed casuistry especially – to offer a method, rather than a solution. All Barlow – or any other casuist – could do was to induce reflection, illustrating the pros and cons of the arguments on any issue. In relation to the doubts that afflicted the Oxford gentlewoman (as also Boyle), Barlow could have recourse to the rhetoric of evangelical Christianity. In other cases, as with the gentleman on whose charitable giving he was asked to give advice, he could only provide overall principles, repeatedly stating how decisions on such matters were 'solely left to the prudence and conscience of every particular Christian; who (after a serious Consideration of all Circumstances), is (hic et nunc [here and now]) to apply the generall Rules of nature and scripture, and soe reduce them to practice'.[90] Similarly, in his paper for Seymour Bowman on the sinfulness of our thoughts, Barlow averred that it was 'A Question not onely difficult, but impossible to be particularly answered by me, or any body else': instead, he offered rules by which, using our God-given ability and duty to assess such matters, 'we can know (as a Carpenter does his timber) whether our waies are straight or crooked'.[91] Hence, at the end of the day, Barlow may not have been able to help Boyle all that much, even if he may have given him useful food for thought (for instance) in his lengthy discussion of the nature and typology of doubt, or in his analysis of charity and its role.[92] It is therefore not surprising that Boyle had to return to his ecclesiastical advisors for further advice in 1691 (the fact that he now went to Burnet and Stillingfleet was probably due to Barlow's increasing infirmity, rather than to any dissatisfaction on Boyle's part). Neither is it surprising that he continued to present his natural philosophical ideas in as convoluted a form as ever. Arguably, Boyle's position in casuistry was symptomatic of his intellectual personality: there was a kind of symbiosis between them, and only a marginal 'influence'. To try to go further is counterproductive.

---

[90]   QC MS 275, item 1, pp. 4–5.
[91]   QC MS 289, p. 161 (item 4, p. 1).
[92]   For the latter, see QC MS 294, pp. 94ff. See also pp. 116–17 above concerning QC MS 275, item 6.

ROBERTVS BOYLE NOBILIS ANGLVS

Plate 1   Portrait of Boyle from the 1680 edition of his writings produced
           by the Geneva publisher, Samuel de Tournes (see below, p. 158).
           This re-engraving of William Faithorne's well-known portrait of
           Boyle by François Diodati was probably the first image of Boyle to
           receive wide circulation.

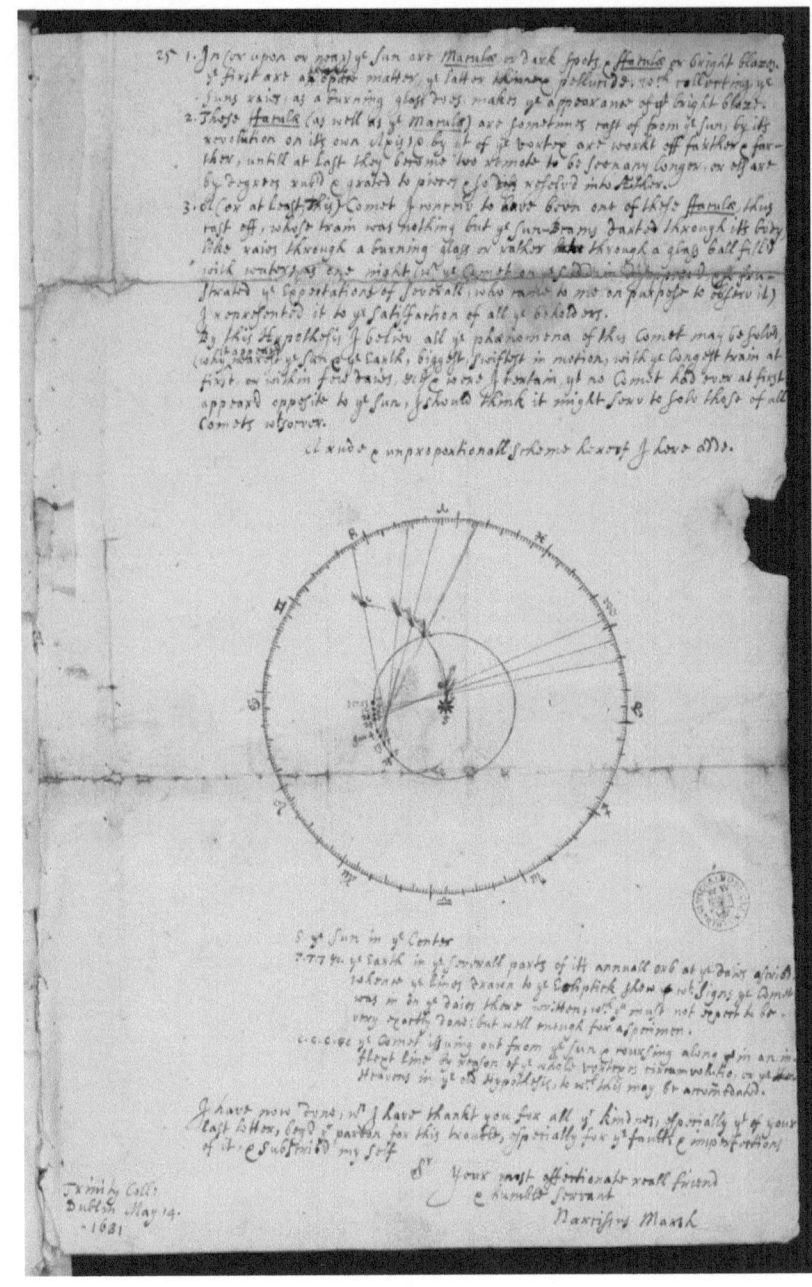

Plate 2   Narcissus Marsh's diagram illustrating his 'Hypothesis of Comets'
(see above, p. 100). Bodleian Library, Oxford, MS Smith 45, p. 25.

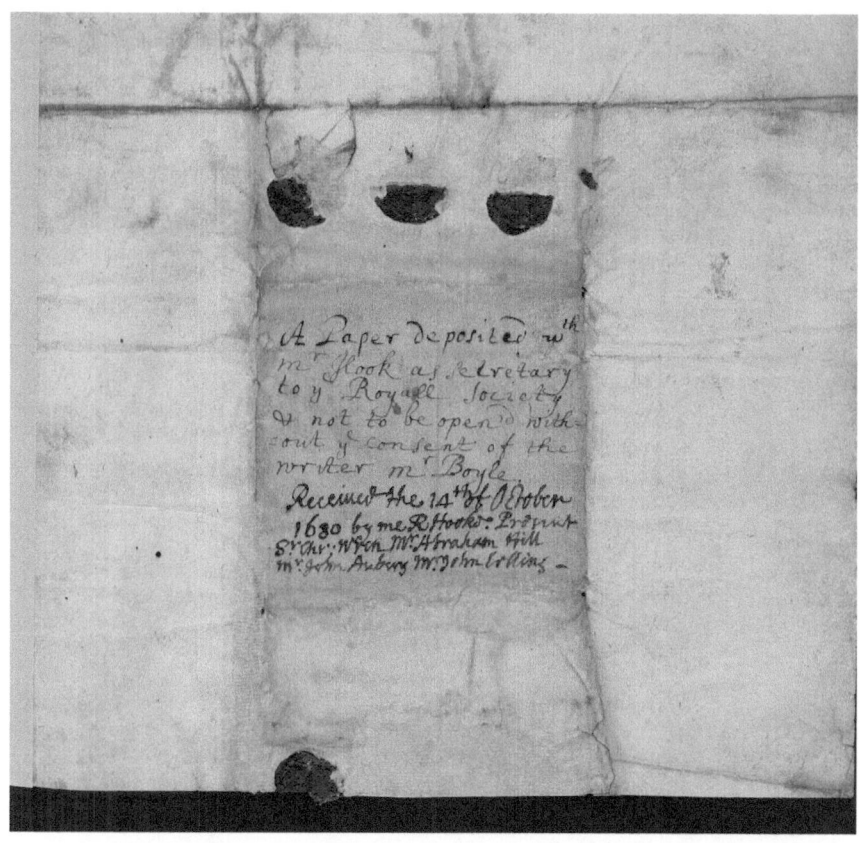

Plate 3   Cover sheet of the document deposited by Boyle at the Royal Society on 14 October 1680 describing the method of making phosphorus from human urine, with an inscription by one of Boyle's amanuenses and an endorsement by Robert Hooke noting his receipt of it and the names of those present, and with the remains of three different seals (see below, p. 147). Royal Society Classified Papers 11 (1) 21.

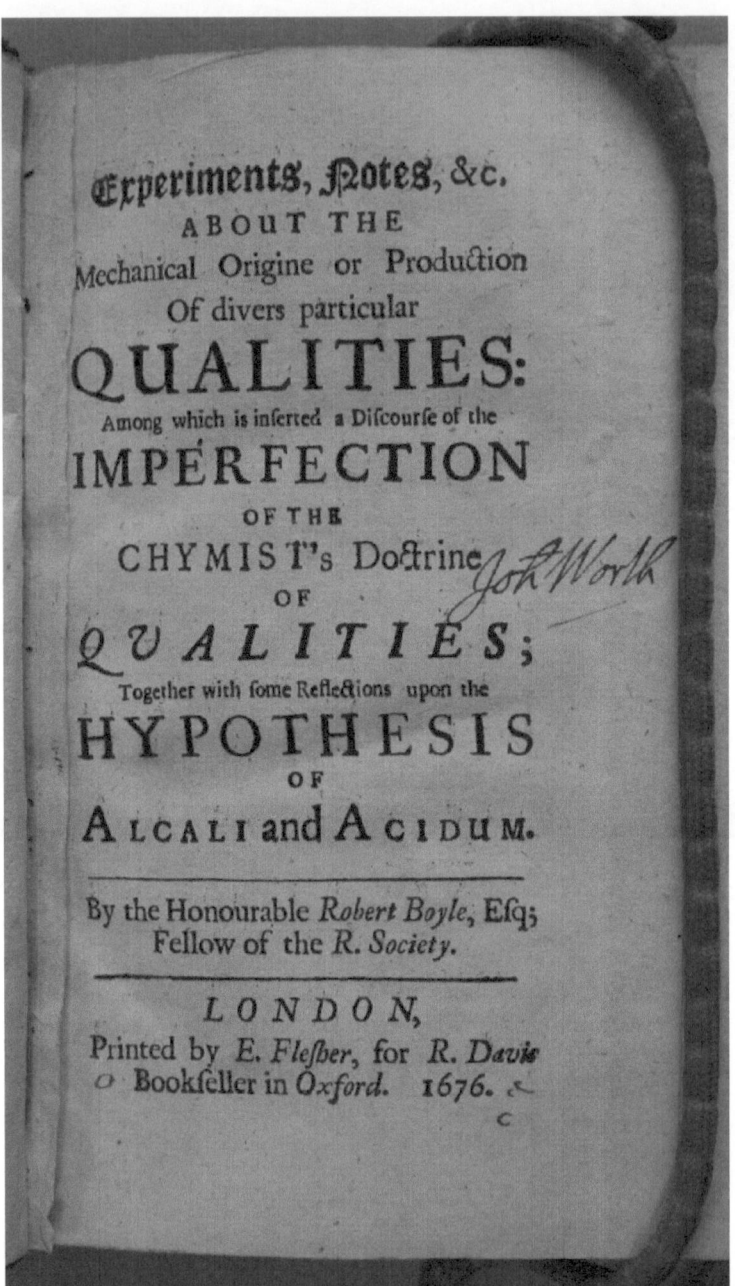

Experiments, Notes, &c.
ABOUT THE
Mechanical Origine or Production
Of divers particular
QUALITIES:
Among which is inserted a Discourse of the
IMPERFECTION
OF THE
CHYMIST's Doctrine
OF
QVALITIES;
Together with some Reflections upon the
HYPOTHESIS
OF
ALCALI and ACIDUM.

By the Honourable *Robert Boyle*, Esq;
Fellow of the *R. Society*.

LONDON,
Printed by E. *Flesher*, for R. *Davis*
Bookseller in *Oxford*. 1676.

Plate 4   Title-page to Boyle's *Experiments, Notes &c. about the Mechanical Origin of Qualities* (1675–76). Edward Worth Library, Dublin.

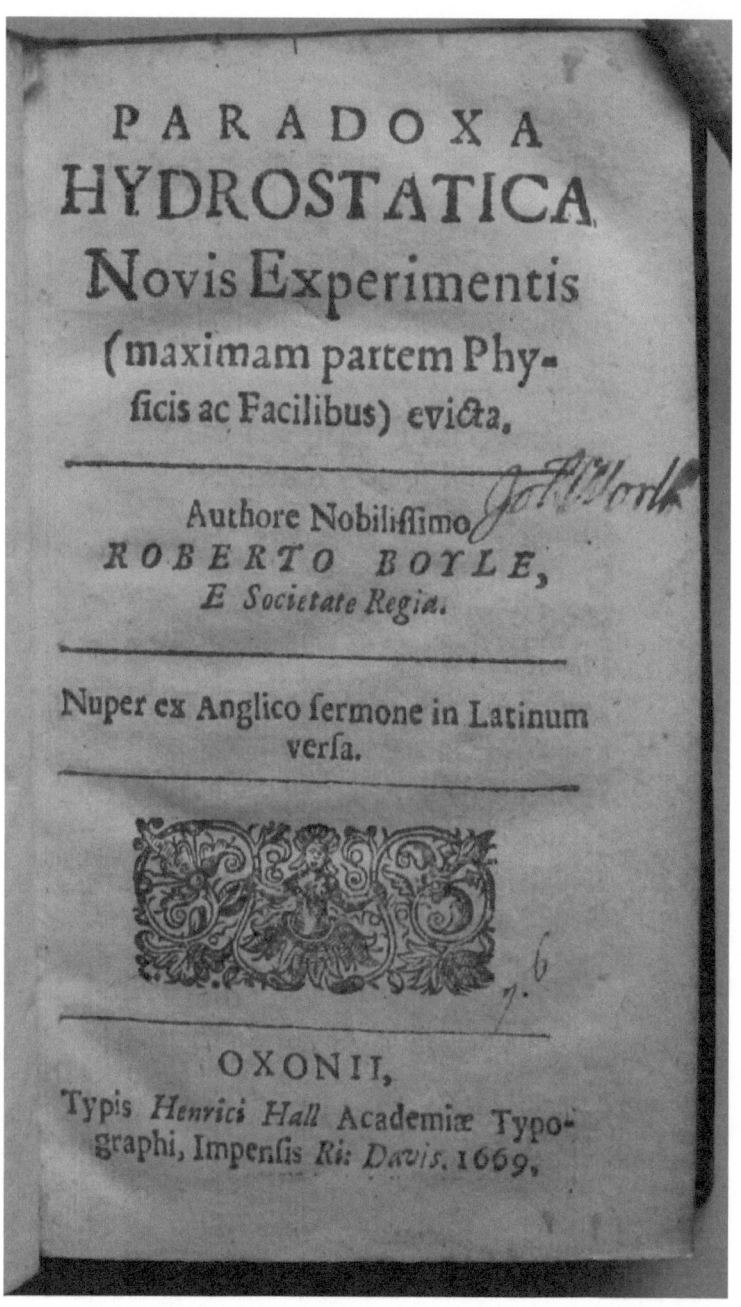

Plate 5   Title-page to the Latin edition of Boyle's *Hydrostatical Paradoxes* (Oxford, 1669), translated by Henry Oldenburg. Edward Worth Library, Dublin.

Scripta Philosophica † *à Nobi-*
lissimo Authore in Lucem edita.

Experimenta nova Physico-mechanica de
Aere; *edita mense Junio,* 16 0.
    Tentamina Physiologica pro datis oc-
casionibus diversis conscripta ; *in lucem*
edita Mense Martio, 1661. *cum historia Fluiditatis*
& Firmitatis adjecta
    Chymicus scepticus ; *mense Augusto editus;*
1661.
    Doctrinæ circa Elaterium et pondus Aeris,
contra objectiones Franc. Linii *Vindicatio; tem-*
pore verno Edita , 1662. *& Thomæ Hobbii*
    Liber de usu Experimentalis Philosophiæ; *edi-*
tus mense Junio, 1663.
    Experimenta nova et Observationes circa Fri-
gus; *sive* Experimentalis Historia Frigoris , in-
choata. *Cui adjicitur* Examen Antiperistaseωs ,
nec non Doctrinæ *Hobbianæ* circa Frigus ; *edita*
mense martio 1665.
    Experimenta et considerata circa Colores ,
mense Maio Edita, 1664.
    Tractatus de Origine Formarum et Qualita-
tum, (juxta Philosophiã Atomicam) consideratis
et Experimentis illustratâ , *mense Aprili editus,*
1666. Huic *accessit* dissertatio de Formis sub.
ordinatis, *mense Septembri,* 1667.

———————————————————————

† *Theologica sunt alterius loci.*

Continuatio

Plate 6   The list of Boyle's writings in *Paradoxa hydrostatica*, sig. B8. As
noted on p. 160, below, this has manuscript emendations which
exactly match those found in all but one other known copy of this
edition. Edward Worth Library, Dublin.

The Right Honoᵇˡᵉ Edward Lord Mountague Viscount Hinchingbrooke Earle of Sandwich Kᵗ. of the most noble Order of the Garter, one of his Maᵗˢ most Honoᵇˡᵉ Privy Counsell, Captaine Generall of the Narrow Seas, Vice-Admirall of England, and Grand Master of the great Wardrobe.

Plate 7   Engraved portrait of Edward Mountagu, First Earl of Sandwich (1625–72), by Abraham Blooteling after Sir Peter Lely. Sandwich was typical of those whose interviews with Boyle are recorded in Workdiary 21 and are discussed in Chapter 9.

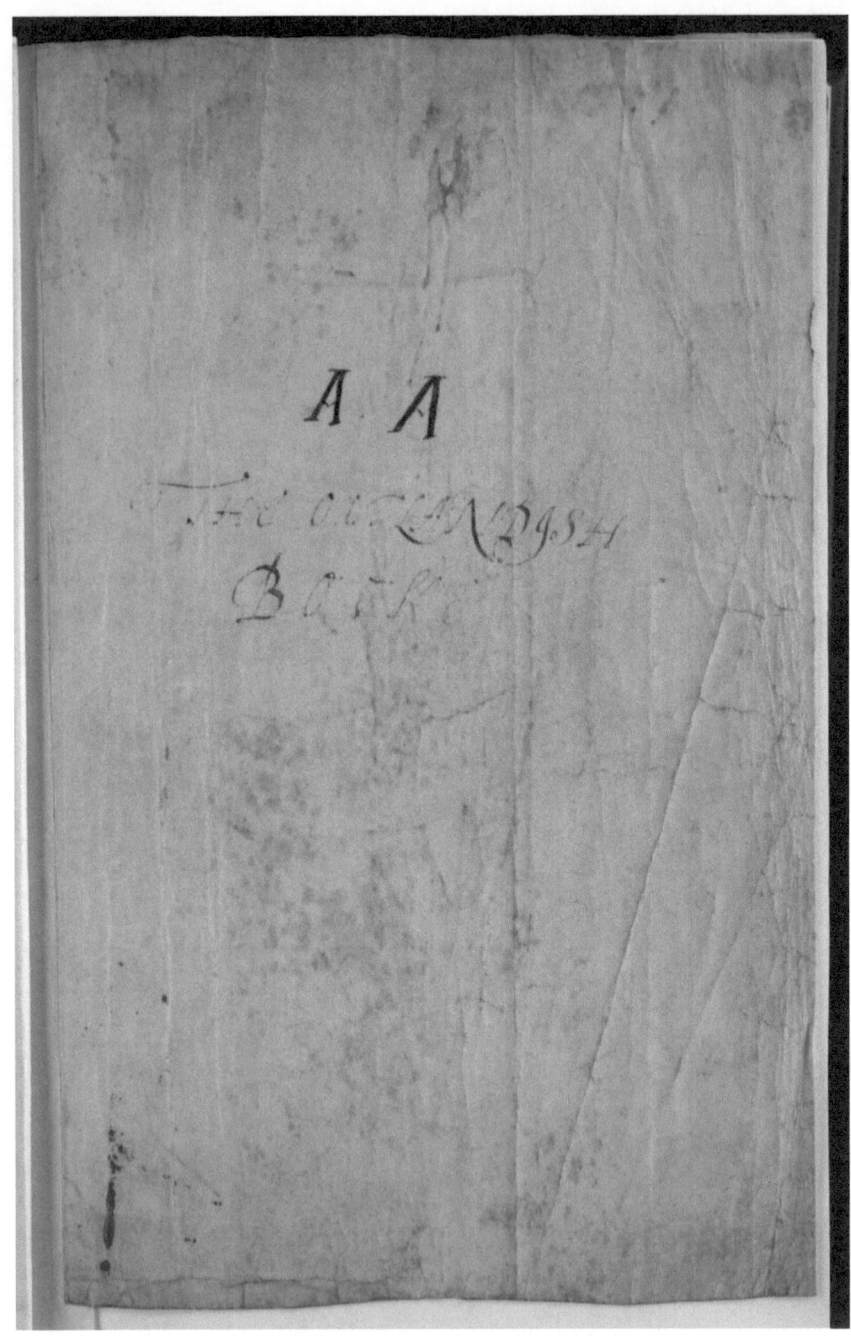

Plate 8   The vellum cover of 'The Outlandish Booke', Workdiary 21. Royal Society Boyle Papers, vol. 27, p. 1.

Chapter 6
Boyle and Secrecy[1]

Our image of Boyle has changed markedly in recent years. This is especially true in relation to alchemy, since, thanks particularly to the work of Lawrence M. Principe, we are aware that Boyle was more fully involved in alchemical pursuits than previous historians had either realised or been happy to accept.[2] We also now have a better understanding of Boyle's complex personality, in which anxious soul-searching co-existed with acute concern about the esteem in which he was held by others and about his relations with the wider world.[3] But, though Boyle was to some extent a victim of traits of the latter kind, he was also a manipulator who could use these and other facets of his personality and status to his advantage, which further complicates matters. The result is that, in the twenty-first century, Boyle has become a more mixed-up and perhaps therefore more interesting figure than the rather lifeless lay saint depicted in the traditional historiography.

On the other hand, though such hitherto neglected facets of Boyle have complicated the traditional view of him, many elements of that view have endured scrutiny. In particular, we need to do justice to the programme in natural philosophy which he so influentially expounded in such writings as *Certain Physiological Essays* (1661), including his insistence on the need for a full account of experimental and other findings as a basis for improved understanding of the natural world, and his championing of the mechanical philosophy. All this formed part of a project for the progress of science both in its own right and for its potential religious and utilitarian spin-offs which Boyle powerfully promoted in his profuse writings over many years. The challenge is to give an account of Boyle which does justice both to the aspects of him that have long been familiar and to the complications introduced by recent scholarship.

Boyle's attitude to secrecy is a key test case in this respect, since the relationship to the secrets tradition of as pivotal figure in the science of his generation as Boyle is obviously worth understanding in its own right. Boyle is all the more interesting because of the extent to which he spans the various contexts in which

---

[1]   Previously published as 'Robert Boyle and Secrecy', in Elaine Leong and Alisha Rankin (eds), *Secrets and Knowledge in Medicine and Science 1500–1800* (Farnham: Ashgate, 2011), pp. 87–104.

[2]   See esp. L.M. Principe, *The Aspiring Adept: Robert Boyle and his Alchemical Quest* (Princeton, NJ, 1998).

[3]   See *Scrupulosity and Science* and *Between God and Science*.

secrecy typically arose at the time – from the fully fledged alchemical tradition in which he participated, with its elaborate protocols of secrecy, encryption and mystification, to the craft tradition and its subset in the form of medical practice, where there was an incentive to keep trade secrets because practitioners' livelihoods depended on them. In Boyle's case, a related issue on which he had strong views concerned authorship and intellectual property, since he believed that the originator of a discovery deserved proper credit for it. There is also the context of virtuoso values, in which secrecy overlapped with the analogous concept of 'rarity', phenomena being valued for their wonderful and inexplicable qualities, and their ownership seen as a mark of exclusivity.[4]

What is today probably Boyle's best-known writing on secrets is one which has in fact only been generally known since 1950, when it was reprinted by Margaret Rowbottom: since then, however, this has often been seen to epitomise the call for openness which was the hallmark of the Baconian scientific tradition of his day.[5] The work in question is Boyle's 'Invitation to a free and generous Communication of Secrets and Receits in Physick', published in *Chymical, Medicinal, and Chyrurgical Addresses: Made to Samuel Hartlib, Esquire* in 1655. In it, Boyle made an impassioned plea for the free dissemination of knowledge, especially medical, answering a range of objections put forward by those who advocated the witholding of useful recipes on the grounds of the prestige that accrued to things for their rarity, or the danger that they might be spoiled in the course of being made common. Boyle's response was that those in possession of secrets that might be of general benefit were obliged on both religious and moral grounds freely to divulge them. He took the view that, even when people had accepted a recipe on condition of secrecy, their duty was to violate this.[6]

Ironically, we now know that this work was compiled at a date which precedes Boyle's hands-on experience of natural philosophy. As we saw in Chapter 2, in his early adult years in the late 1640s, Boyle was surprisingly little interested in science, instead writing treatises on ethical theory and related texts which he hoped would help to reform the morals of his benighted peers. The 'Invitation' forms part of this group of writings by Boyle and it fits into this tradition in its essentially moralistic stance: though it clearly reflects Boyle's experience as a consumer of medications, it is arguably rather limited in its sympathy for the views of their producers.

---

[4]   For a helpful recent overview of virtuoso values, see Brian Cowan, *The Social Life of Coffee: The Emergence of the British Coffeehouse* (New Haven, CT and London, 2005), esp. pp. 10ff. On Boyle and intellectual property, see *Scrupulosity and Science*, esp. pp. 219–21.

[5]   See, for example, William Eamon, 'From Secrets of Nature to Public Knowledge', in D.C. Lindberg and R.S. Westman (eds), *Reappraisals of the Scientific Revolution* (Cambridge, 1990), pp. 333–65, on p. 351, and Jan Golinski, 'Chemistry in the Scientific Revolution: Problems of Language and Communication', in ibid., pp. 367–96, on p. 383.

[6]   *Works*, vol. 1, pp. 1–12.

Secrecy in Boyle's Early Natural Philosophy

When we look at Boyle's attitudes in the period *after* he discovered natural philosophy in 1649, we find him displaying typical virtuoso attitudes, including, perhaps not surprisingly, the classic language of secrecy. This is seen in the workdiaries he compiled in these years, which provide a rather breathless account of his pursuit of medical recipes and other material from fellow enthusiasts, and which often evaluate the information he obtained in such terms. Thus Boyle explained how his newly made acquaintance Sir Kenelm Digby 'commends above all his Secrets' a formula which he had obtained from Basil Valentine's *Currus Triumphalis*, while of another of Digby's recipes, Boyle noted: 'This he tells me is his Greatest secret in perfuming.'[7]

Much the same is true of Boyle's letters in these years, for instance one to his Oxford acquaintance Ralph Bathurst in 1656, in which, acknowledging receipt of a recipe for distilling spirit of roses, Boyle wrote: 'The secresy that you command in reference to the receipt, its owne nature would have exacted from me.'[8] The notes about Boyle in Samuel Hartlib's 'Ephemerides' tell a similar story. Thus Hartlib recorded how Boyle accounted a recipe for essence of jasmine 'one of the greatest secrets', while, in telling of a friend of Boyle's who had a method of making apples grow unusually large, he added: 'but Mr Boyle is bound to secrecy and binds others to it'.[9]

By this time, Boyle had been exposed to an apprenticeship both in high-level chemical experimentation and in the secrets tradition by one of the classic exponents of both, George Starkey, the American alchemist who came to England in 1650 and almost immediately struck up a friendship with Boyle. Starkey's profuse letters to Boyle in the year after they met are full of news of secret processes he had discovered which he vouchsafed to Boyle only on condition that he kept them to himself. In fact, we now know that Boyle was spectacularly bad at doing so, that (in the words of William R. Newman and Lawrence M. Principe) he 'appears to have been a constant "leak" of privileged information'.[10] Thus in 1653 we find him vouchsafing one of Starkey's choicest secrets – for the preparation of wine from fermented vegetables – to another of his chymical friends, Frederick Clodius, Hartlib's son-in-law, with the rider: 'Having received this Processe from a freind as a Secret; I shall beg you would

---

[7] Workdiary 12–5, 13–6.

[8] *Correspondence*, vol. 1, p. 204. The original manuscript of this letter has recently come to light among Bathurst's papers at Trinity College, Oxford, Fellows 2/1/1. I am grateful to Philip Beeley for alerting me to it.

[9] Hartlib Papers, University of Sheffield, 29/7/16A, 29/8/11B.

[10] W.R. Newman and L.M. Principe, *Alchemy Tried in the Fire* (Chicago, IL, 2002), p. 265, and see ibid., pp. 265–7.

not let it loose that Name.'[11] This formula does not suggest that Boyle was reverting to the ethos of openness he had advocated in his 'Invitation'. Rather, his behaviour seems to combine a formal acceptance of the protocols involved with a certain casualness about observing them, possibly as if he considered that his superior social status placed him above such constraints.

A comparable ambivalence is in evidence in Boyle's practice concerning the recipe known as *ens veneris*, a copper compound with powerful medicinal qualities which he and Starkey were inspired to prepare on the basis of a joint reading of J.B. van Helmont.[12] Increasingly, Boyle took credit for this nostrum, alluding only casually to Starkey's role in its discovery, but his attitude towards its dissemination fluctuated. His protégé Henry Oldenburg obtained 'the substance of the processe' from Boyle in 1659 while on his Continental travels with Boyle's nephew Richard Jones, and he almost immediately offered it to the Dresden chemist Gansland. In doing so, however, he specifically stated that the originator of the nostrum, Boyle, had made it a condition 'that I communicate it to no one, but such as have first faithfully promised not to reveal this secret, reserving it for their own use and practice only'.[13] Boyle himself, on the other hand, clearly purveyed the medication widely to those who needed it, while in 1663 he published full details of it in his *Usefulness of Natural Philosophy*.[14]

A further curious episode of about the same period involves an exchange of letters between Boyle and that classic exponent of genteel virtuoso values John Evelyn. Evelyn had sent Boyle a recipe for varnish, but he explained his ambivalence about publishing such recipes lest he 'debase much of their esteeme, by prostituting them to the Vulgar'. In his reply, Boyle expressed the hope 'that some Expedient may be found to reconcile the disclosure of many secrets, with the keeping up & secureing the Reputation of Learning'. His wording is slightly puzzling. In part, it echoed Evelyn's, who had outlined a plan for an academy at which well-born youth would be schooled in virtuoso values '(not without an Oath of seacresy)', and the superintendent of which would from time to time divulge specimens of such material 'for the reputation of Learning, and benefit of the Nation'. But Boyle's take seems perceptibly different, reflecting his developing agenda for demarcating between 'experimentall & notionall Learning', and his view that the best way to encourage his fellow countrymen to value the former was to disseminate 'reall & usefull Productions' which exemplified its potential.[15]

---

[11]   *Correspondence*, vol. 1, p. 149.

[12]   See W.R. Newman, *Gehennical Fire* (Cambridge, MA, 1994), pp. 71–2, and Newman and Principe, *Alchemy Tried in the Fire*, pp. 9, 221–2.

[13]   *Correspondence*, vol. 1, pp. 328–9; A.R. and M.B. Hall (eds), *The Correspondence of Henry Oldenburg* (13 vols, Madison, WI and London, 1965–86), vol. 1, pp. 243–4. I owe the latter reference to David Haycock.

[14]   *Works*, vol. 3, pp. 391–2, 500–505.

[15]   *Correspondence*, vol. 1, pp. 212–15.

By this time, we have reached the period in the late 1650s when Boyle developed the distinctive outlook in natural philosophy for which he has since been remembered, writing a notable series of natural philosophical treatises that were published in the 1660s and which established him as one of the most influential exponents of the 'new' empirical science of his period. In these, as in his comment to Evelyn, we find echoes of the ethos of his early 'Invitation to Communication', and it is important to remember that Boyle never lost the moral imperative of his early writings, simply transferring it to new fields, particularly the study of nature. On the other hand, it was supplemented by new criteria that reflected the natural philosophical agenda which he was expounding at this time and which had been missing before.

Boyle's Statements on Secrecy

What, therefore, did Boyle say about secrecy in the writings he compiled at this point, in which he laid down clear protocols for the pursuit of science, in terms both of its explanatory goals and of the most appropriate methods to achieve them? Boyle included various statements on the issue which we must consider before turning to the evidence of Boyle's actual practice in relation to secrets in these and other writings.

First, there is *The Sceptical Chymist* (1661) and Boyle's familiar strictures on the mode of writing associated with the chemists whom he attacked in that work. He wrote how:

> though much may be said to Excuse the Chymists when they write Darkly, and Ænigmatically, about the Preparation of their *Elixir*, and Some few other grand *Arcana*, the divulging of which they may upon Grounds Plausible enough esteem unfit; yet when they pretend to teach the General Principles of Natural Philosophers, this Equivocall Way of Writing is not to be endur'd. For in such Speculative Enquiries, where the naked Knowledge of the Truth is the thing Principally aim'd at, what does he teach me worth thanks that does not, if he can, make his Notion intelligible to me, but by Mystical Termes, and Ambiguous Phrases darkens what he should clear up; and makes me add the Trouble of guessing at the sence of what he Equivocally expresses, to that of examining the Truth of what he seems to deliver. And if the matter of the Philosophers Stone, and the manner of preparing it, be such Mysteries as they would have the World believe them, they may Write Intelligibly and Clearly of the Principles of mixt Bodies in General, without Discovering what they call the Great Work.[16]

---

[16] *Works*, vol. 2, p. 292.

As will be seen, despite Boyle's acknowledgement that there were certain secrets it was perfectly legitimate for chemical writers not to divulge, he was equally emphatic that openness and clarity were appropriate in relation to natural philosophy, and elsewhere he was to express disdain for those who possessed knowledge but '(Chymist like) keep it secret'.[17]

A second relevant statement appears in the 'Proemial Essay' to *Certain Physiological Essays* (1661), where Boyle addressed the issue of 'why in the ensuing Essays I have mention'd divers Experiments which I have not plainly and circumstantially enough delivered'.[18] In this case, Boyle's response relates to secrecy in the context of craft practices: after first apologising that he might have been unduly succinct in expounding some processes, he went on to explain how he had 'purposely omitted some manual Circumstances', justifying this partly out of solicitude for tradesmen whose livelihood depended on such products, and partly by demarcating between what was appropriate in natural philosophy and what was appropriate in other fields in a manner which echoes *The Sceptical Chymist*. Even more pertinent are Boyle's further considerations:

> Thirdly, I mention'd some things but darkly, either because I receiv'd them upon Condition of secrecy, or because some ingenious persons that communicated them to me, or others to whom I imparted them, do yet make, and need to make, a pecuniary advantage of them. Fourthly, And some things that, either having been the fruits of my own Labours, or obtain'd in Exchange of such, are freely at my own disposal, I have not yet thought fit so plainly to reveal, not out of an envious design of having them bury'd with me, but that I may be always provided with some Rarity to barter with those Secretists that will not part with one Secret but in Exchange for another, and think nothing worth their desiring that is known already to above one or two Persons. And I think it very lawful to reserve always some conceal'd Experiments by me, wherewith to obtain the secrets of others, which being thereby gain'd, the other (as being no longer necessary to the former end) may freely be communicated.

It is interesting how the last phrase actually echoes the earlier 'Free Invitation', as does his disavowal concerning secrets of 'an envious design of having them bury'd with me'. Equally notable is the reference to his practice of 'bartering' recipes: this bears out the note of the physician and memorialist John Ward on a visit to Boyle in 1667 that he 'hardly parts with any mony to buy Receipts, but only exchanges', and a similar line was later taken by the physician and naturalist Martin Lister, who told Oldenburg that he considered such exchange 'the most

---

[17]   Ibid., vol. 5, p. 418.
[18]   Ibid., vol. 2, pp. 30–31.

creditable & proper use of secretts'.[19] This looks forward to Boyle's actual practice in ways I will come to shortly, the implication being that Boyle 'concealed' experiments only in order to swap them for others and thus make it possible freely to communicate both – though it might be felt that this itself undercut the ethos of secrecy, since the effect of such practice would be to devalue the very item Boyle had just bartered.

A further reflection on related issues occurs in Boyle's 'Preamble' to Part 2, section 2 of *The Usefulness of Natural Philosophy*, mainly written in the late 1650s although not published with the 'First Tome' of that work in 1663, instead seeing the light of day only in 1671.[20] This passage is interesting not least because Boyle here shows his awareness of the books of secrets tradition, actually alluding to authors like 'Cardan, Weckar, and Baptista Porta'. In it, Boyle sought to answer the 'objection, whether I doe not injure Tradesmen by discovering so plainly those things, which our Laws call the Mysteries of their Arts', claiming firstly that 'I never divulge *all* the Secrets and practices necessary to the exercise of any one Trade, contenting my self to deliver here and there upon occasion some few particular Experiments, that make for my present purpose'. Secondly, he argued that 'it is not the Custome of Tradesmen to buy Books', whereas those who typically purchased these were ill-equipped to implement the technical prescriptions that they contained, not least since this usually required 'a Manuall dexterity' which was 'not to be learnt from Bookes, but to be obtain by imitation and use'. But he went on to claim that:

> though some little inconvenience may happen to some Tradesmen by the disclosing some of their Experiments to practicall Naturalists, yet that may be more than compensated, partly, by what may be contributed to the perfecting of such experiments themselves, and, partly by the diffused Knowledge and sagacity of Philosophers, and by those new Inventions, which may probably be expected from such persons, especially if they be furnished with Variety of hints from the practices already in use. For these Inventions of ingenious heads doe, when once grown into request, set many Mechanical hands a worke, and supply Tradesmen with new meanes of getting a livelyhood or even inriching themselves.

Boyle then proceeded to give examples of the ways tradesmen might benefit from the superior inventions which naturalists produced, instancing the profitable trade in optical instruments which London instrument-makers like Richard Reeve enjoyed on the basis of scientific improvements going back to Galileo.

---

19   Folger Shakespeare Library, Washington, DC, MS V.a.296, fol. 20; *Correspondence of Oldenburg*, vol. 10, p. 383.
20   *Works*, vol. 6, pp. 397ff. For parallel passages, see ibid., vol. 4, pp. 114, 175, 180.

Again, this is a revealing passage, notable for its condescension towards humble operatives in comparison with the lofty status accorded to natural philosophers on the basis of the analytical understanding to which they aspired. What was the source of this? In part, it stemmed from the strong faith in the potential of experimental philosophy to provide a true understanding of nature which formed so influential a part of Boyle's philosophical outlook. But where did *this* come from? Here we may perhaps discern the extent to which Boyle transposed to his role as a natural philosopher the attitudes associated with his privileged status as one of the richest aristocrats in England, including the instinctive expectation of deference from his social inferiors. Boyle displayed similar attitudes in his encounter with the world of projecting when he became involved in a scheme for desalinising sea-water in the 1680s and when he seems similarly to have presumed that his ability to understand the rationale of processes gave him a right simply to ignore the interests of rivals.[21]

A final general reflection by Boyle on the pros and cons of divulging secrets had appeared in the first section of Part 2 of *The Usefulness of Natural Philosophy*, published in 1663, where, prior to giving the recipe for the medicine 'Pilulæ Lunares', he apologised for not naming the 'very Ancient and experience'd Chymist' from whom he had derived it, on the grounds that it was often 'more prejudicial then gratefull to one that makes an advantage by the Practise of Physick, to annex in his lifetime his name to some of his Receipts or Processes', since this encouraged people to circumvent him in various ways.[22] This opened up a whole can of worms about divulging 'communicated recipes' in medicine, Boyle's convolutions concerning which I have expounded elsewhere.[23] Overall, however, it will be seen that, through such programmatic statements, Boyle was provided with a range of justifications on the one hand for divulging secrets, but on the other for being somewhat circumspect in doing so. Indeed, it could be argued that he was thus provided with the potential for a considerable amount of flexibility in practice, enabling him to justify a range of strategies as it suited him. What, therefore, did he actually do?

Boyle's Attitude in Practice

It is certainly the case that, in Boyle's time as earlier, there was a clear expectation fostered by the secrets tradition that it would have been legitimate for Boyle to keep things secret had he so wished. This is seen in those of his correspondents

---

[21]   See *Between God and Science*, p. 216. For background on the role of status see Steven Shapin, *A Social History of Truth* (Chicago, IL, 1994).

[22]   *Works*, vol. 3, pp. 485–6.

[23]   *Scrupulosity and Science*, ch. 9.

who presumed that he might be reluctant to divulge details of recipes, asking for information only 'if not kept as a secret', or promising that 'you may engage me to secrecy in what you please'; Henry Oldenburg similarly assured Boyle concerning data he received from him that 'you may be sure, I shall keep them secret, as long as I receive no leave to divulge them'.[24] Boyle was certainly capable of withholding information when he wanted to, as part of a repertoire of authorial apologies of which his works are full and which illustrate a perceptible degree of self-indulgence on his part.[25] Thus he even occasionally offered to supply missing information by word of mouth, while he often excused himself for failing to publish matters in full because he was not yet satisfied with his investigations or on other comparable grounds, in one case explaining more enigmatically how 'cogent considerations forbid me at present to publish' a process.[26]

Boyle also commonly invoked the ethos of secrecy on the grounds that he had been given details of a practice on condition that he did not divulge its rationale. Thus, in connection with a method of staining marble red, he explained how this had been revealed to him 'upon condition of secresy (which I have to this day inviolably kept)'. Elsewhere, he tells of 'a strange way of preserving Fruits, whereby even *Goos-berries* have been kept for many Moneths, without the addition of Sugar, Salt, or other tangible Bodies', owned by 'an eminent Naturalist, a Friend of yours and mine', adding: 'but all that I dare yet tell you, is, That he assures me his Secret consists in a new and artificial way of keeping them from the Air'.[27] In at least some such cases, it is quite possible that this ostensible solicitousness for Boyle's informants was in fact a subterfuge for protecting secrets of Boyle's own.

In terms of Boyle's actual practice in relation to secrets, on the other hand, the very cases cited above are revealing. In the first of them, Boyle explained how he helped the 'ingenious Stone-cutter in *Oxford*' whose secret it was how 'to improve his Invention by making it practicable with other Colours than Red', thus illustrating an element of interchange that seems to have characterised many such transactions on Boyle's part. The second, on the other hand, has an element of provisionality about it, implying that the status of the secret was quite contingent and that Boyle's ambition was to demystify it.

Indeed, in many ways the latter instance does seem to provide the key to Boyle's attitude on such matters. Since he saw the goal of natural philosophy as to furnish explanations, the ultimate challenge was to make all secrets unmysterious. The same is equally true of phenomena he described as 'rarities', where Boyle is to be found taking a similar attitude, in this case sometimes

---

[24] *Correspondence*, vol. 3, pp. 259, 344; vol. 4, p. 121.

[25] See *Scrupulosity and Science*, ch. 7.

[26] *Works*, vol. 3, p. 510; vol. 7, p. 383. See also ibid., p. 387; and, for example, vol. 2, p. 373; vol. 3, p. 412; vol. 5, p. 407; vol. 10, p. 382.

[27] Ibid., vol. 3, p. 358; vol. 10, p. 139.

destroying items treasured by connoisseurs in order to elucidate their physical properties.[28] In the case of secrets, Boyle took it for granted that they all could ultimately be 'cracked', either by obtaining information from the person whose secret it was, or by investigations of his own, and on various occasions he gives interesting hints as to how he went about this. Sometimes, as in the case just cited, he gives a kind of provisional account of a secret, implying that in time he would be in a position to elucidate it, while elsewhere he gives examples of secrets which he obtained by exchange and then de-mystified. Thus he tells of a secret he obtained from:

> a not unlearned Emperick, who was exceedingly cry'd up for the Cures he did, especially in difficult Distempers of the Brain, by a certain Remedy, which he call'd sometimes his *Aurum Potabile*, and sometimes his *Panacæa*; and having obtain'd from this Man, in exchange of a Chymical Secret of mine he was greedy of, the way of making this so celebrated Medicine, I found that the main thing in it was the Spirit of Soot, drawn after a somewhat unusual, but not excellent manner; in which Spirit, Flowers of *Sulphur* were, by a certain way, brought to be dissolv'd, and swim in little drops that look'd of a golden colour.[29]

On other occasions, it seems likely that Boyle conducted his investigations by a mixture of inquiry and experiment. Concerning one such 'choice Secret', he explained how:

> though I could not learn a considerable Particular or two, which belong to the Delicacy of it; yet (partly by putting Questions, and partly by some Tryals of my own) I attain'd to the substance of this Mystery, as they call it, which seems to be this.

He then went on to outline how, by the application of a coating of wax to an exemplar written in a special ink, fine lettering could be produced on an engraved plate.[30] Perhaps the most interesting example of this mixture of interrogation and experimentation concerns the noctiluca, the phosphorescent substances brought to the attention of Boyle and other virtuosi of the Royal Society by J.D. Krafft and others from 1677 onwards and which Boyle wrote up in his *The Aerial Noctiluca* (1680). This was a typical virtuoso display which capitalised on its strangeness and wonder, while Krafft and the other originators

---

[28]   See, for example, ibid., vol. 7, pp. 25–6.

[29]   Ibid. vol. 3, p. 384. For other references to the use of exchange, see, for example, ibid., p. 385, vol. 13, p. 248.

[30]   Ibid., vol. 6, pp. 499–500.

of the substance were careful to protect their rights by keeping the method of preparing it secret.[31] Boyle, however, tried his usual methods. He explained how:

> having at Mr. *Krafft's* desire, imparted to him somewhat that I discover'd about uncommon *Mercuries*, (which I had then communicated but to one Person in the World) he, in requital, confest to me at parting, that at least the principal matter of his *Phosphorus's*, was somewhat that belong'd to the Body of Man. This intimation, though but very general, was therefore very welcome to me, because, though I have often thought it probable, that a *shining substance* may, by Spagyrical Art, be obtain'd from more kinds of Bodies than one: yet designing, in the first place, to try if I could hit upon such a *Phosphorus* as I saw was preparable, the Advertisement sav'd me (for some time) the labor of ranging among various Bodies, and directed me to exercise my industry in a narrower compass.[32]

Boyle thereupon pursued this line of inquiry, leading him to focus on urine as the bodily substance from which phosphorus could be made, but he reached an impasse because he was unable to work out the temperature at which the conglomerate needed to be heated to separate out the impurities it contained. He went on:

> However, adhering to the first choice I had made of a *fit* matter, I did not desist to work upon it by the ways I judg'd the most hopeful, when a learned and ingenious Stranger, (*A. G. M. D.* Countreyman, if I mistake not, to Mr. *Krafft*) who had newly made an Excursion into *England*, to see the Countrey, having, in a Visit he was pleas'd to make me, occasionally discoursed, among other things, about the *German Noctiluca*, whereof he soon perceiv'd I knew the true matter, and had wrought much upon it. He said something about the degree of Fire, that made me afterwards think, when I reflected on it, that *that* was the only thing I wanted to succeed in my endeavors.[33]

In other words, Boyle found that, by combining hints he was given by their proprietors with his own expertise, he could almost invariably get to the bottom of any secret that presented itself.

Hence, perhaps we are coming back by a different route to the ethos of Boyle's early 'Invitation to Free Communication'. Boyle seems to have it taken for granted that secrecy was a contingent status which it was appropriate to try

---

[31]   For an account of the whole episode, see Jan Golinski, 'A Noble Spectacle: Phosphorus and the Public Cultures of Science in the Early Royal Society', *Isis*, 80 (1989), 11–39. See also J.R. Partington, 'The Early History of Phosphorus', *Science Progress*, 30 (1936), 402–12.

[32]   *Works*, vol. 9, p. 273.

[33]   Ibid., pp. 273–4. This was probably Ambrose Godfrey Hanckwitz, later Boyle's operator.

to break down through investigation by one means or other. In a number of cases, his sense also becomes apparent that the demands of natural philosophy and humanity, particularly for secrets which might be of great public benefit, overruled the selfish interests of those who might otherwise have taken their nostrums to the grave with them. On one occasion, he himself acted as the intermediary in this, when he was able to divulge a method of increasing the yield of corn which he obtained from its originator, 'who was pleas'd to make choice of me to instruct his Secret with, that in case he dyed before me, the publick might not loose it'.[34] On others he failed, telling wistfully of excellent nostrums that were lost to posterity because of the selfishness of their proprietor, as in an early workdiary record from Boyle's Irish colleague Gerard Boate, who informed him of a 'Bohemian Physitian, (whose name I cannot readily remember,)' who had an infallible cure for all sorts of agues which worked within 24 hours: 'he added that this Secret dy'd with the Inventor'.[35]

What can only be described as a similar moralism is in evidence in certain cases where Boyle took the view that inventions or phenomena were best *not* divulged, since these were often powerful processes which he believed could have a harmful effect if placed in the wrong hands, and which therefore 'my love of Mankind has oblig'd me to conceal, even from my nearest Friends'.[36] Indeed, this sense of a threat to the good of mankind is virtually the only grounds on which Boyle states that he was permanently withholding information. He took a similar line concerning a technique which could be used for counterfeiting coins: 'the fear of teaching bad men a skill that probably they will not otherwise acquire, makes me forbear to mention it'.[37]

A similar consideration potentially arose in relation to alchemy. Thus, in a manuscript set of headings for and against divulging the arcanum to which I first drew attention and which William Newman has since published in full, Boyle cited as an argument against such revelation: 'That it would much disorder the affairs of Mankind, Favour Tyranny, and bring a general Confusion, turning the World topsy turvy'.[38] This was echoed by Isaac Newton in the letter he was stimulated to write to Henry Oldenburg by the article Boyle published in *Philosophical Transactions* in 1676 about the incalescence of mercury, in which he criticised the author for publicly divulging matters which 'may possibly be

---

[34]   Ibid., vol. 6, p. 417.

[35]   Workdiary 6–14. Cf. *Works*, vol. 3, pp. 337, 449, 538.

[36]   Ibid., vol. 9, p. 281.

[37]   Ibid., vol. 8, pp. 538–9. For further comparable references, see ibid., vol. 6, p. 430; vol. 9, p. 293; MS 187, fol. 166.

[38]   *Scrupulosity and Science*, p. 114; Newman, *Gehennical Fire*, pp. 254–5. Since such arguments are balanced by contrary ones, the overall import of the document is ambiguous.

an inlet to something more noble, not to be communicated without immense dammage to the world if there should be any verity in the Hermetick writers'.[39]

Secrecy and Alchemy

Turning, therefore, to Boyle's concern with alchemy, this was clearly something of a special case – hardly surprisingly in view of the fact that, even in the passage in *The Sceptical Chymist* already cited in which Boyle criticised the obscurity of chymists' language, he agreed that it was legitimate for the higher arcana to be withheld. In fact, this is an area where Boyle *was* circumspect. A good example is provided by a letter he wrote in March 1680 to the New England physician William Avery, who had been inspired by Boyle's reputation to make contact with him. Initially, they discussed medical matters, including the efficacy of herbs found in the New World but not the Old, but when Avery plucked up courage to ask Boyle for information about the Helmontian alkahest, he received the following reply:

> I perceive you have had a very partial representation of my skill in Chymistry & Physick, to the latter of which I do not pretend, nor to the former any further than as an innocent Recreation & an Instrument to discover the nature of things. Therefore you much misaddress yourselfe when you would learn of me the preparation of such high Arcana as the Alkahest, which I never in any booke of mine ownd my selfe to be a Possessor of, and which I conceive to be so lyable to be misemployed to dangerous purposes, that if I were master of the Alkahest & the way of makeing it, I should thinke my selfe oblig'd not to communicate it to any for whom a long acquaintance had not given me a particular esteeme & friendship[.] And to add that upon the by, I doubt not but Helmonts sayes true where he declares that the Alkahest is tædiosissimae Præparationis; & that other Arcana majora of the Chymical Philosophers, tho the processes were deliverd without Riddles, would be very difficult to prepare by persons not already <very> well versd both in the more common, & some of the more subtile operations of Chymistry. Thô I do not therefore blame your aspireing to the highest secrets, yet I am sorry to see no more ground than I do to expect that either by my means or other mens you will easily be made master of them.[40]

It is interesting how Boyle makes a virtue of the fact that he had never made claims to such secrets in his books, as also how he invokes moral scruples about

---

[39] H.W. Turnbull et al. (eds), *The Correspondence of Isaac Newton* (7 vols, Cambridge, 1959–77), vol. 2, p. 2.

[40] *Correspondence*, vol. 5, pp. 191–3.

divulging so potentially dangerous a formula (except to those he trusted), and questions the viability of the process. But it is also worth observing that we know from Lawrence Principe's researches that this disavowal on Boyle's part was highly disingenuous, in that he *had* pursued exactly the knowledge which he here effects to disclaim.[41] Indeed, in his later years Boyle made a collection of just the kind of secrets to which Avery evidently aspired, and, though the collection itself is lost, Boyle's prefatory letter to it survives, in which he wrote how:

> since I find myself now grown old; I think it time to comply with my former intentions, to leave a kind of Hermetick Legacy to the Studious Disciples of that Art, & to deliver candidly in the annex'd Paper, some Processes Chymical & Medicinal, that are less simple and plain than those barely Luciferous ones, I have been wont to affect; and of a more difficult & elaborate kind, than those I have hitherto publish'd, and more of kin to the noblest Hermetick Secrets, or as Helmont styles them, *Arcana majora*. Some of these I have made & try'd; Others I have, (thô not without much difficulty) obtain'd by Exchange or Otherwise from those that <affirm they> knew them to be real, & were themselves competent Judges; as being some of them Disciples of true Adepts, <or> otherwise admitted to their acquaintance & conversation. Most of these Processes are clearly enough deliver'd, and of the rest there is plainly set down without deceitful terms, as much as may serve to make what is literally taught, to be of great utility: thô the full & compleat uses are not mention'd, *partly* because in spite of my Philanthropy, I was ingag'd to secrecy, as to some of these uses, and *partly* because (I must ingeniously confess it) I am not yet, or perhaps ever shall be, acquainted with them my self.[42]

Paradoxically, as death approached, Boyle seems to have aspired to at least a partial openness even in this field, though within this very passage he reverts to such commonplace excuses as the need to protect the secrets of others, or the fact that his knowledge was incomplete.

Earlier, on the other hand, Principe has illustrated how Boyle used standard alchemical techniques of encryption, including word substitution and dispersal, to obscure the meaning of alchemical material he recorded. A minor element of this appears even in published writings such as *The Usefulness of Natural Philosophy*, though this usually takes the form of withholding crucial information and seeding the text with phrases and formulae which would alert alchemical readers to the significance of the processes he was expounding.[43] In

---

[41]   Principe, *Aspiring Adept*, esp. ch. 5.

[42]   *Works*, vol. 12, pp. 365–6. For related documents, see Principe, *Aspiring Adept*, pp. 300ff.

[43]   See L.M. Principe, 'Robert Boyle's Alchemical Secrecy: Codes, Ciphers and Concealments', *Ambix*, 39 (1992), 63–74; Principe, *Aspiring Adept*, pp. 143ff.

his manuscripts, however, he was much more overt, not least in using strange word substitutions, as in the following passage in one of his workdiaries which reads in the original as follows:

> With Dicla and Mardium prepare Nigerus by 7 operations, then mix with it Barakid a 12th part, keep it about 3 dayes in digestion till opening the Vessell from time to time toward the latter end, you perceive almost all the Barakid to have gaind the lower part of the Mixture.

(The fifth word from the end is annotated by Boyle: 'quaere, if not the Upper'.) What this actually means is:

> With Cinnabar and Iron filings prepare Mercury by 7 operations, then mix with it Silver a 12th part, keep it about 3 days in digestion till opening the vessel from time to time toward the latter end, you perceive almost all the Silver to have gained the lower [upper] part of the Mixture.[44]

Since such substitutions were often used in his laboratory notes, they were evidently partly intended to protect his data from his own assistants, and in this connection an intriguing piece of evidence survives in the form of an oath which Boyle evidently drafted with a view to having his assistants affirm it (interestingly, his own scruples about oaths made him refrain from making them take a vow). The formula was:

> I do hereby solemnly & faithfully promise & ingage myself that ... I wil not knowingly discover to any person whatsoever, whether directly or indirectly, any process, medicine, or other experiment, which he [Boyle] shal injoin me to keep secret & not impart; without his consent first obtain'd to communicate it.[45]

Hence, here we do see Boyle invoking secrecy, even if it might be felt that this was legitimate enough in relation to his own personal information management.

Secrecy and Institutionalisation

However, perhaps the oddest example of Boyle's ambivalence over secrecy is to be seen in one facet of his relationship with the Royal Society, thus adding to the complexities of understanding the relationship between secrecy and

---

[44] Workdiary 34–55. Cf. *Boyle Papers*, p. 147.
[45] Royal Society MS 189, fol. 13, reproduced in Shapin, *Social History of Truth*, p. 404: see also ibid., p. 403, and *Scrupulosity and Science*, p. 67n.

modernising trends in the science of his day as exemplified by new institutions like the Royal Society. In the 1660s, Boyle had been in the forefront of initiatives to use institutional deposits with the newly founded body as a means by which Fellows with 'a philosophical notion or invention, not yet made out' could have it sealed up 'till it could be perfected, and so brought to light'.[46] The rationale was that 'this might be allowed for the better securing inventions to their authors' – in other words, the protection of intellectual property rights – and it was evidently in this connection that Boyle availed himself of the facility in the 1660s. Subsequently, he reverted to the practice, and his later deposits were opened at a meeting of the society after his death, on 9 February 1692. Moreover, since the documents involved were described at the time as 'Arcana', there is a clear link to the secrets tradition.[47] The papers in question do indeed turn out to record recipes of just the kind that others might have protected as 'secrets', but there is a strange contrast between them which encapsulates Boyle's residual ambivalence in this respect.

One paper, dated 30 November 1683 and deposited by Boyle later that year in sealed form, outlined an effective method of testing the salinity of water using a solution of silver nitrate, a key tool in relation to the patent application for desalinising sea-water with which he was involved at that point.[48] Interestingly, this was not the only institutional deposit that occurred in connection with the desalination project. An announcement in the *London Gazette* for 23 January 1685 reveals that various of the recipes it deployed were deposited with the Lord Mayor of London, 'lest a Secret of so great importance to the Publick might come to be lost, if lodged only in the knowledge of a few Persons therein concerned'.[49] It is revealing that the term 'secret' was used in this public statement, though it is significant that the concern was to ensure that the information in question was preserved – a rather Boylean concept, as we have seen. As for the paper deposited with the Royal Society, this was even more 'public' in its rationale, since although deposited in manuscript form, it was in no way an 'arcanum', being presented as ready for print, with elaborate instructions as to how shoulder notes

---

[46]   Ibid., pp. 219–20.

[47]   Royal Society Copy Council Minutes, vol. 2, pp. 113–14, quoted in *Works*, vol. 12, p. xxv. See also ibid., pp. xxv–xxvi.

[48]   Royal Society Classified Papers, 6, 51. For the context, see *Between God and Science*, pp. 215ff., esp. p. 218.

[49]   Quoted in R.E.W. Maddison, 'Studies in the Life of Robert Boyle F.R.S.: Part II. Salt Water Freshened', *Notes and Records of the Royal Society*, 9 (1952), 196–216, on p. 207. There is no reason to think that Boyle was personally responsible for this (the account of the deposit in the *London Gazette* suggests that it was the King who took the initiative; the process deposited with the Royal Society does not seem to have been included). See also ibid., pp. 204–7. For a different reading of this episode, see Adrian Johns, *Piracy: The Intellectual Property Wars from Gutenberg to Gates* (Chicago, IL, 2009), pp. 73ff.

should be placed and the like which were exactly followed when it appeared in *Philosophical Transactions* in 1693.[50]

However, the second item, deposited on 14 October 1680, is rather bizarre.[51] It describes a method of preparing phosphorus from human urine which is evidently the very process which, as we saw earlier, Boyle had discovered by a typical combination of hints from informants and experimentation. What is extraordinary about it is the manner in which it was deposited. It was carefully folded up and secured with three seals, one of them Boyle's own and the others evidently those of two of the four witnesses to the deposit (Plate 3). It really does *feel* like a 'secret' carefully salted away, even if the excitement of the occasion when it was opened in 1692 was rather reduced when Boyle's former assistant, Frederick Slare, observed that chemists had since discovered 'a more convenient & shorter way of preparing' the substance (he might have added that the information contained in the paper had in fact been in print since later in 1680, when Boyle had included an almost identical version of 'The Process' at the end of his *Aerial Noctiluca*).[52]

It is perhaps ironic that Boyle arguably came closest to the secrets tradition in the institutional setting where he should have been furthest from it. After all, his deposits with the Royal Society were of details of processes similar to those which their proprietors often refused to disclose, and which Boyle would therefore typically have sought to demystify, as we have seen. Just as it seemed as if – alchemy aside – Boyle displayed a rather modern attitude, he pulls us up short. Indeed, this makes it hardly surprising that to Boyle's successors his attitude towards secrecy often seemed frustratingly inconsistent and unclear. Thus, in reporting to Locke after Boyle's death about their communication on alchemical matters, Newton complained how Boyle had shown a certain 'reservedness' concerning a secret that interested him, suspecting that Boyle had deliberately withheld part of it from him.[53] Similarly, Samuel Cottereaux Duclos of the Parisian Académie des Sciences was critical of Boyle not least for his refusal to divulge key particulars in chemical reactions, which he thought

---

[50] *Works*, vol. 12, pp. xxviii and 165–74. It is perhaps worth noting that the third paper opened at the same time was a copy of Boyle's queries concerning mineral waters, and it is unlikely that he saw this as an 'arcanum': ibid., p. xxvi, and vol. 10, p. xxx.

[51] Classified Papers 11 (1) 21, dated 30 September 1680. For the deposit, see Hooke's report in Birch, *Royal Society*, vol. 4, pp. 55–6.

[52] Copy Council Minutes, vol. 2, pp. 113–14. Compare *Works*, vol. 9, p. 303, and vol. 12, pp. 163–4 (this overlap is not there commented on: this will be rectified in corrigenda to the edition). See also ibid., vol. 9, p. 300, for Boyle's report on his deposit of this item at the Royal Society. For a commentary, see Partington, 'Early History of Phosphorus', esp. p. 407.

[53] See *Correspondence of Newton*, vol. 3, pp. 218. See also ibid., pp. 193, 195, 216–19.

showed an unhelpful double standard on Boyle's part.[54] There seems little doubt that Boyle's equivocation on such matters had a significant effect on his legacy, meaning that it was more mixed than might otherwise have been the case. In all, Boyle seems likely to remain a somewhat paradoxical figure.

---

[54]   See Victor D. Boantza, 'Chymical Philosophy and Boyle's Incongruous Philosophical Chymistry', in Ofer Gal and Raz Chen-Morris (eds), *Science in the Age of Baroque* (Dordrecht, 2012), pp. 257–84. It is perhaps worth noting here a similar attitude on the part of an early annotater of the 1663 edition of Tome 1 of Boyle's *Usefulness of Natural Philosophy*, evidently Algernon Capel, Second Earl of Essex, whose bookplate is in the copy of the book in question, Bodleian Library, Oxford, 12 Θ 1329 (the hand is similar to that of letters from Essex in British Library Add. MS 40629, fols 199, 209, 217). He gave vent to his frustration at Boyle's ambivalence about divulging information in a series of sarcastic comments, for instance 'thank you, Sir', where Boyle wrote: 'what I can further say of this matter, I must not declare in this place' (p. 198 [*sic*]; *Works*, vol. 3, p. 412 – a typical example of the kind of generalised excuse referred to on p. 139 above). For other annotations see pp. 137–8, 140–41, 146, 170, 175, 177, 181, 183, 185, 188–9, 193, 196, 354. I am indebted to Will Poole for drawing my attention to this volume.

Chapter 7
Boyle and the Uses of Print[1]

This chapter was originally presented as part of a celebration of the 275th anniversary of the Edward Worth Library in Dublin that took place in 2008. It was appropriate that a paper about Boyle should form part of those proceedings, not least since Boyle was born in Ireland and Irish estates provided the bulk of his income all his life, so that Thomas Duddy has aptly described him as 'Irish by privilege'.[2] On the other hand, after leaving the country at the age of eight and a half Boyle thereafter only once briefly visited it, and to contemporaries he was very much 'the *English Philosopher*', the leading natural philosopher of his generation, and one of the emblems of the Royal Society in its early years.[3] As a prolific author, his works formed one of the staples of libraries in his own generation and the next, and it is not surprising to find his books well represented in the Worth Library, some collected by John Worth, Dean of St Patrick's Cathedral, and some by his son Edward, a Dublin physician. Indeed, copies of many of the books cited in this chapter are to be found in the library, while its illustrations are taken from volumes in that collection.

What I want to do in this chapter is to use these books to reflect more generally on Boyle's attitude to print, which was more complex and interesting than might perhaps have been expected. Boyle had a high view of the significance of the printing press, like his mentor Francis Bacon, for whom printing, along with gunpowder and the compass, were the three inventions which had 'altered the whole face and state of things' since classical antiquity.[4] Boyle's own sense of the crucial yet normative role of print is perhaps most clearly shown in a treatise he wrote on medical remedies in his later years, in which he took the view that such cures should be as freely available as possible, regardless of any threat this

---

[1]   Previously published as 'Robert Boyle and the Uses of Print', in Danielle Westerhof (ed.), *The Alchemy of Medicine and Print: The Edward Worth Library, Dublin* (Dublin: Four Courts Press, 2010), pp. 110–24.

[2]   For this quotation, and for comments on Boyle's 'hybrid identity', see Thomas Duddy, 'Introduction' to *Dictionary of Irish Philosophers* (Bristol, 2004), pp. vii–xvi, on pp. xi–xii.

[3]   David Abercromby, *A Discourse of Wit* (London, 1685), p. 200. See also *Works*, vol. 1, p. lxxv; vol. 10, pp. lviii, 357.

[4]   Bacon, *Novum organum*, i, 129, in Graham Rees with Maria Wakely (eds), *The Instauration magna Part II: Novum organum and Associated Texts* (Oxford, 2004), pp. 194–5.

might present to a monopolistic medical profession. To reinforce the point, he continued caustically:

> We would have lacked the benefit of printing if the printing-shops had been removed for fear lest they should reduce the profit of so many scribes, who sought their livelihood from the transcription of books (and perhaps piled up considerable wealth for themselves in so doing). What would this be? It would seem to smack of nothing less than barbarism in Europe! For, in most of the territories of the Turks, and in most of the regions of the Orient, printing-shops have either been eliminated or were not allowed to set themselves up, merely to gratify the calligraphers.[5]

Clearly, for him it was axiomatic that the way in which print made possible the wide dissemination of ideas was highly desirable.[6] Equally important was its role in preserving ideas that might otherwise be lost, as was stressed in the preface to Boyle's *Of the Usefulness of Natural Philosophy* (1663) by his 'publisher' Robert Sharrock, who bewailed the insecurity of writings which survived 'only in single, or at most in few written Copies'. Sharrock illustrated the point by instancing the loss of one of the essays in this very book of Boyle's, going on to note how printing might have preserved such works from classical antiquity as Solomon's 'Universal History of Vegetables' or Democritus' book of experiments, both of which were instead irrevocably lost.[7]

In fact, Boyle was slightly slower to discover the value of print than he might have been, since when he first began to write prolifically on natural philosophy at Oxford in the late 1650s he did indeed entrust his ideas to a limited number of written copies which he circulated among his friends.[8] But at the end of that decade and the start of the new one – in juxtaposition with the Stuart Restoration – he suddenly became aware of the value of print, and espoused the new medium with the zeal of a convert. Between 1660 and 1666 he published 11 books, representing an average of 140,000 words per year. These included the first 'tome' of his *Usefulness of Natural Philosophy* (1663), together with many of the works for which he is best known, including *New Experiments Physico-Mechanicall, Touching the Spring of the Air and its Effects* (1660) and two controversial works stemming from this; *Certain Physiological Essays* (1661); *The Origin of Forms and Qualities* (1666–67); his experimental histories of cold and colours (1664–65), and perhaps his most famous work of all, *The Sceptical*

---

5   *Works*, vol. 14, pp. 290–91. The work in question survives only in a translation made for Boyle into Latin, so this is a modern translation.

6   For other comments by Boyle on printing, see ibid., vol. 6, pp. 396, 423.

7   Ibid., vol. 3, pp. 193–4.

8   For evidence of the circulation of *Certain Physiological Essays* thus, see R.G. Frank, *Harvey and the Oxford Physiologists* (Berkeley, CA and Los Angeles, CA, 1980), p. 123.

Chymist (1661). All this had been preceded by a devotional work, *Some Motives and Incentives to the Love of God*, usually known as *Seraphic Love* (1659), which in fact outsold all Boyle's scientific books, while it is perhaps worth noting that prior to this, in 1658–59, Boyle had been involved in publishing English editions of texts of interest to him, and such surrogate publications perhaps acted as a prelude to the burst of publication on his part which followed.[9]

Boyle continued to publish extensively for the rest of his life. Although he never again equalled the peak of such activity that he achieved in the early 1660s, he produced a steady stream of books on both experimental and theological topics over the next two and half decades, including such key works as *Experiments, Notes &c. about the Mechanical Origin of Qualities* (1675–76) (Plate 4), *A Free Enquiry into the Vulgarly Received Notion of Nature* (1686) and *The Christian Virtuoso* (1690–91). By the time of his death some forty treatises by him had appeared in print, to be followed by a handful more shortly afterwards: in total, these make up 12 volumes, each averaging over 500 pages, in the edition of *The Works of Robert Boyle* published in 1999–2000. He also championed the printing press as an agent of dissemination for evangelistic purposes, not least by supporting the publication of first the New Testament and then the Old Testament in Gaelic in the 1680s, for distribution in Ireland and in the Scottish Highlands. Indeed, this even necessitated the cutting of a special font of Irish type for the purpose, which was executed at Boyle's expense, some punches from which still survive.[10]

Concerning the publication of Boyle's own writings, a good deal of evidence survives illustrating his solicitousness about how his works were presented in print. He tended not to deal directly with printers, but to work through a 'publisher' – initially Robert Sharrock, whom I have already mentioned; later Henry Oldenburg, first Secretary of the Royal Society; and, following Oldenburg's death in 1677, Boyle's trusted servant, John Warr. In the case of both Sharrock and Oldenburg, details are sometimes preserved in extant letters between them and Boyle as to the choices that had to be made – concerning the appropriate length of a volume, for instance, or whether the best format for a book would be quarto or octavo.[11] In addition, Boyle seems to have evolved a canny strategy for keeping printers on their toes by never entrusting all of his works to a single printing house. Throughout the 1660s and 1670s, some of his works were printed in London and some at Oxford, which seems to have been a strategic decision rather than one based on his own whereabouts, since this often

[9]   *Works*, vol. 1, p. xlii; *Between God and Science*, ch. 8.
[10]   See Chapter 4 in this volume, and *Between God and Science*, pp. 197–9, 227.
[11]   *Works*, vol. 1, pp. xlviff. For discussion of the bulk of a book, see ibid., vol. 8, p. xvi. For the use of quarto as against octavo, see ibid., vol. 1, p. cxxxi, and *Correspondence*, vol. 1, pp. 437–8.

resulted in publishing arrangements having to be made by proxy. Moreover, even after he more or less abandoned Oxford printers from the early 1680s onwards, he still continued a 'dual-track' policy by patronising more than one London printer at the same time.[12]

Perhaps the clearest evidence of Boyle's conscious strategy concerning his books concerns the provision of Latin translations of them. Though he could obviously read Latin fluently, Boyle invariably wrote in English: all his extant drafts are in the vernacular, and it was from such texts that his books were set in type to produce the editions I have itemised above. But he was acutely aware that the language of international intellectual communication remained Latin. Indeed, Continental savants sometimes complained about the policy of Boyle and other early Fellows of the Royal Society of publishing in English. This was also the case with the successful journal Oldenburg launched in 1665, *Philosophical Transactions*, which caused similar problems for those 'clumsy in the English language'.[13] Boyle realised that, if his works were to be read as widely as he would like and thus to enjoy the international influence he thought they deserved, he needed to have them translated into Latin. Hence, from the point when he began to publish his scientific writings in English, he also started to have them translated into Latin for the Continental market (interestingly, he seems to have been less assiduous about having his theological works translated, evidently presuming that the market for them was primarily an English-speaking one[14]). So important did he see this process of translation that he sometimes began it in parallel with the publication of the English edition, with the result that, insofar as Boyle made last-minute additions to the English version of a book while it was at the press, these are missing from the Latin one.[15] Moreover he sometimes added material to the Latin editions which did not appear in the English ones, particularly material intended for his Continental readers: thus, in the case of the Latin translation of his *Defence* of his *New Experiments touching the Spring of the Air*, he added a riposte to the critique of his work by a Dutch scholar, Anthony Deusing, which he never printed in English.[16]

We learn quite a lot about the preparation of these translations from Boyle's correspondence, especially with Robert Sharrock and other Oxford figures who undertook such work before it was increasingly taken over by Oldenburg. As Sharrock pointed out, the kind of topics Boyle wrote about were rather unfamiliar to the Oxford scholars who assisted him, who were more accustomed

---

[12]   *Works*, vol. 1, pp. xlixff.

[13]   Quoted in Michael Hunter, *Science and Society in Restoration England* (Cambridge, 1981), p. 53.

[14]   *Works*, vol. 1, pp. lxii–lxvii.

[15]   Ibid., vol. 1, p. lx.

[16]   Ibid., vol. 3, pp. xiv–xv, 6–8n. The Worth Library has a copy of Deusing's *Considerationes* (1662), though not of Boyle's response.

to 'the Vulgar Philosophy, Humanity & Divinity'. They also had difficulty finding words for some of the technical terms Boyle used in his scientific writings: how was 'quicksilver' best rendered in Latin, for instance, while even the concept of 'virtuosi' presented a challenge.[17] Indeed, partly because of the difficulties he encountered with such scholastic translators – which came to a climax with the Latin translation of *The Origin of Forms and Qualities*, which was generally agreed to be rather unsatisfactory[18] – from the late 1660s such figures were ousted by Henry Oldenburg. After Oldenburg's death, Boyle evidently considered reverting to Oxford figures, but was again dissatisfied; instead, he deputed translation to such protégés of his as David Abercromby, a lapsed Jesuit who worked for Boyle in the 1680s, and who translated four of his books into Latin, including *Memoirs for the Natural History of Human Blood* (1684) and *A Free Enquiry into the Vulgarly Received Notion of Nature* (1686). Indeed, Abercromby even accompanied his Latin versions of Boyle's books with promotional puffs about the author, acclaiming Boyle's experimental expertise in *Human Blood*, while describing *Notion of Nature* as 'the new system of a new philosophy which fundamentally overthrows the foundation – namely Nature – of all views hitherto held in philosophical matters'.[19]

Physically, perhaps the strangest Latin translation of a work by Boyle was the version of his paper on the incalescence of mercury published in *Philosophical Transactions* in 1676, entitled 'An Experimental Discourse of *Quicksilver* growing hot with *Gold*' and attributed to one 'B.R.', a rather crude pseudonym for Boyle.[20] For this covertly alchemical text was presented in Latin and English in two parallel columns, a format never previously used in the journal and never used again, the initiative for deploying which must therefore have been Boyle's. The object was clearly to provide a version for the journal's normal English readers, but also to make it available to the international cognoscenti to whose attention Boyle wished to bring this arcane process; potentially, this was the key to transmuting base metals into gold, and Boyle was therefore keen for fellow adepts to contact him to share their findings on the subject. How far he succeeded in the latter aim is unfortunately unclear, but his openness with such a significant secret aroused the ire of his younger contemporary and fellow alchemist Isaac Newton, who (as we saw in Chapter 6) wrote to Oldenburg warning the author not to meddle in matters which might represent 'an inlet to

---

[17]   Ibid., vol. 1, pp. lxix–lxx, lxxiii–lxxiv.

[18]   Ibid., vol. 1, p. lxxi; vol. 5, pp. xxix–xxx.

[19]   Ibid., vol. 1, p. lxxii; vol. 10, pp. 98–101 (on p. 100) and 572–6 (on p. 575).

[20]   Ibid., vol. 8, pp. 553ff. I have used the title from the contents list of this issue of *Philosophical Transactions* as a whole.

something more noble', the communication of which could be very dangerous 'if there should be any verity in the Hermetick writers'.[21]

The idiosyncratic use of print on this occasion, however, was typical of Boyle. Indeed, there is considerable evidence for his seeing the potential for print to be used in slightly abnormal ways to suit his purposes, which is significant in itself and which is worth exploring here. The first example concerns the nature of the books Boyle published in the early 1670s. As already indicated, Boyle published massively from the early to mid-1660s; thereafter, however, there was something of a pause, during which he published a number of scientific papers in Oldenburg's newly established *Philosophical Transactions*. In the course of this, Boyle seems to have been one of the first to come to relish the format of the journal article, which enabled him to set out his findings in a briefer, more provisional and, if appropriate, more speculative way than would have been the case with a full-length book. Hence, when he returned to publishing volumes of his writings from 1670 onwards, he started to adopt a new format for them which bore some relation to his earlier journal articles, in that the works in question comprised several separate short pieces juxtaposed: the link to his earlier experience is suggested by the fact that one of the works in question is actually dedicated to Oldenburg as publisher of *Philosophical Transactions*.[22]

One of these publications is so brief that, unbound, it looks like one of the ephemeral pamphlets produced at the time: this was entitled *Tracts written by the Honourable Robert Boyle Of a Discovery of the Admirable Rarefaction of the Air* (1670). More commonly, however, he collected a number of such 'tracts' together, though rather confusingly with a title page starting with the same general title, *Tracts written by the Honourable Robert Boyle*, or *Tracts* followed by a list of titles, with Boyle's details after that. Such volumes appeared in 1670 (this is in addition to the *Rarefaction* one already mentioned), in 1672, in 1673 and in 1674. In fact, at least one contemporary collector thought they should be bound as a series with the general title of *Tracts*.[23] Usually, however, they are distinguished by referring to them by the first main treatise within them, respectively *About the Cosmical Qualities of Things*, *The Relation betwixt Flame and Air*, *Observations about the Saltness of the Sea* and *Suspicions about some Hidden Qualities of the Air*. Each volume comprised a number of such 'tracts', each of them self-contained, and usually with a separate title page and pagination: indeed, this has caused something of a problem for bibliographers, who have been uncertain whether to catalogue the components individually or as

---

[21]   H.W. Turnbull et al. (eds), *The Correspondence of Isaac Newton* (7 vols, Cambridge, 1959–77), vol. 2, pp. 1–2.

[22]   *Works*, vol. 1, pp. xxxvi–xxxviii.

[23]   Ibid., vol. 6, p. xlvn.

groups.[24] Boyle used the same practice for other works of this period, notably the second 'tome' of his *Usefulness of Natural Philosophy* (1671), which comprises six separate essays, the correct ordering of which was indicated only on a discrete contents leaf, and his *Mechanical Origin of Qualities* (1675–76) (Plate 4), again a set of ostensibly self-sufficient tracts each with its own pagination and title page, the intended order of which was given only in an inserted leaf entitled 'Directions for the *Book-binder*.'[25]

The early 1680s saw a publication which is equally revealing in terms of Boyle's attitude to print, his *Memoirs for the Natural History of Human Blood* (1684), on which I have written a paper with Harriet Knight.[26] This is a rather different kind of book by Boyle – a work explicitly inspired by Francis Bacon in which Boyle collected together information about a topic arranged according to classified headings or titles, thus contributing to the great bank of data or 'natural history' which would act as 'the basic stuff and raw material of true induction.'[27] In this case, the work in question has usually been seen by modern scholars as rather disappointing. For one thing, the data it presented had mainly been collected by Boyle over twenty years previously, which meant that it failed to take account of important work on related topics that had been done by scientists in the interim, particularly the demonstration of the micro-structure of blood by the Dutch microscopist Antonie van Leeuwenhoek. Moreover, it did not even classify all the data it contained according to the category to which they belonged.

Yet what is interesting is that this seems to have been deliberate on Boyle's part. He saw this as a provisional publication, a working paper which might stimulate further research and thus contribute to the great natural historical project to which Bacon and he aspired. To some extent, he hoped that others might be encouraged to carry out comparable work, and there is some evidence that they were. But he also himself set to work almost immediately on a second edition of the book, for which various materials survive although it was never published, in which he sought to rectify some of the defects of the first. In part, he did so by collecting new research findings and in part by revising his list of headings to take account both of these and of the discoveries of van Leeuwenhoek and others since his original research had been carried out.[28] Scholars who have judged the

---

[24] This is especially true with *Mechanical Origin of Qualities*, of which each component is separately entered in Wing.

[25] Ibid., vol. 6, pp. liv, lviii, 403; vol. 8, pp. xxxvi, 319.

[26] Harriet Knight and Michael Hunter, 'Robert Boyle's *Memoirs for the Natural History of Human Blood* (1684): Print, Manuscript and the Impact of Baconianism in Seventeenth-century Medical Science', *Medical History*, 51 (2007), 145–64.

[27] Bacon, *Instauratio magna Part II: Novum organum*, pp. 454–5.

[28] See Knight and Hunter, 'Boyle's *Memoirs*', pp. 159ff. The texts in question are published in full in Michael Hunter and Harriet Knight (eds), *Unpublished Material*

published work as Boyle's final word on the subject and damned it accordingly have failed to see the extent to which he was again being rather original in using the medium of print in a flexible, contingent way as a means of advancing knowledge. Indeed, in his 'Post-Script' to the first edition, he placed this in the context of an interestingly progressivist view of science, commenting how:

> I presume, that our enlightned Posterity [will] arrive at such attainments, that the Discoveries and Performances, upon which the present Age most values it self, will appear so easy, or so inconsiderable to them, that they will be tempted to wonder, that things to them so obvious, should lye so long conceal'd to us, or be so much priz'd by us; *whom* they will perhaps look upon with some kind of disdainful Pity, unless they have either the equity to consider, as well the smalness of our helps, as that of our Attainments; or the generous gratitude to remember the Difficulties this Age surmounted, in breaking the Ice, and smoothing the way for them, and thereby contributing to those Advantages, that have enabled them so much to surpass us.[29]

A further way in which Boyle was unusual in his deployment of print was in his use of private printing to give material a circulation that was more extensive than could be achieved in manuscript, but over the distribution of which he had complete control. One such item unfortunately no longer survives at all, namely a printed questionnaire concerning the characteristics of the air which he apparently circulated in preparation for the work published after his death by John Locke as *The General History of the Air* (1692). This was presumably similar to manuscript lists of queries which *are* extant and served a similar function – though, since it is lost, it is impossible to be sure of this.[30] On the other hand, various other such publications *have* just about survived. One is the trial edition Boyle produced in 1688 of 50 medical recipes from a much larger collection of such material which he had accumulated over many years, under the title *Some Receipts of Medicines. For the most part Parable and Simple. Sent to a Friend in America*: the friend in question was Dr William Avery of Boston, Massachusetts, whom we encountered in Chapter 6. This is an extraordinarily rare book: only one copy survived until modern times, which was fortunately microfilmed and made available on Early English Books Online, because that copy – which was in the British Library – can no longer be located. The reason for its rarity is explained by the publication history of the work, which Boyle recounted in manuscript material relating to the preface to the larger collection of recipes

---

Relating to Robert Boyle's 'Memoirs for the Natural History of Human Blood', Robert Boyle Project Occasional Papers No. 2 (London, 2005).

[29] *Works*, vol. 10, pp. 96–7.

[30] Ibid., vol. 12, pp. xi–xii, and see above, p. 65.

which followed the trial version, *Medicinal Experiments*, which came out in three volumes after his death. Boyle explained that he had the book 'printed but not publish'd', buying the whole impression himself, 'without excepting those copies, that are wont to be claimd & taken by those that had to do with the presse'. Copies were distributed 'gratis; not only to physitians, & surgeons, but cheifly to divines & Ladyes, & other persons residing in the countrey that were wont out of charity to give medicins to the poore'. Their copies were provided on condition that they disseminated only individual recipes and reported back on their efficacy; they were not allowed to circulate the book as a whole, so as to prevent the text being pirated. Hence the book's virtual disappearance since.[31]

Boyle produced two other publications in 1688 which are also very rare and which may well have been comparably produced. One of these is stated to have been printed by Edward Jones, who was to become royal printer on the accession of William and Mary the following year, and he may well also have been responsible for *Some Receipts of Medicines*, though no printer's name is given in that volume. This is *A Catalogue of the Philosophical Books and Tracts, Written by the Honourable Robert Boyle Esquire*, printed in Latin in 1688 and then issued in an English translation in 1689.[32] It was a sequel to a series of similar lists of Boyle's writings which had hitherto been appended to his books and about which I will say more shortly, but what is curious about this is that it was a separate 22-page pamphlet – a highly ephemeral publication which is accordingly rare.

The other was a broadside advertisement, in the format of a single-sheet newspaper of the period. This bizarre publication is entitled *An Advertisement of Mr Boyle, about the Loss of many of his Writings*. It comprised a prolonged apology concerning the 'unwelcom Accidents' that had befallen Boyle's manuscripts, which reached a climax when a bottle of concentrated sulphuric acid held by one of his assistants unfortunately broke immediately over a chest of drawers which he had had specially made to house his manuscripts, many of which were ruined, including 'some that I most valu'd ... insomuch that there remain'd not words enough undefac'd to declare what Subjects they concern'd'. Boyle went on to explain how, because of this, he had resolved to publish more of his writings in unfinished form than might otherwise have been the case, and to warn readers of the likely defects that would result. Only three copies of this item survive, all inscribed 'Printed in May 1688' and with a word in the text altered in the hand of one of Boyle's amanuenses.[33] Again, Boyle was using print for bizarre, private purposes of his own.

---

[31]   Ibid., vol. 11, pp. xxvii–xxviii, and 173–86, and *Scrupulosity and Science*, pp. 207–8.
[32]   *Works*, vol. 11, pp. xxxi–xxxiv and 187–97. On Jones, see ibid., vol. 1, pp. liv–lv.
[33]   Ibid., vol. 11, pp. xxiv, 169–71.

The purposes for which he did so overlapped with ones for which he had used the medium in his earlier books, and this was as a kind of loud-hailer – a means of publicly apologising for defects in his publications and, equally important, of asserting his rights as an author and attacking those he believed had infringed these. Perhaps the most striking examples of such practices on Boyle's part were stimulated by the unauthorised issue of a series of reprints of his books by an enterprising Genevan publisher, Samuel de Tournes, from 1676 onwards, so it is with these that I will begin. De Tournes was a regular exhibitor at the German book fairs, and he did a line in reprints of medical and scientific works by eminent authors which served a real need in the international intellectual community of the day. It must therefore have seemed natural for him to extend this treatment to Boyle, and, after initially issuing Latin reprints of individual writings, in 1677 he went further by issuing a volume with a general title page announcing the collection as the *Opera Varia* of Robert Boyle, *Nobilissimi Angli, et Societatis Regiæ Dignissimi Socii*, and by including an apologetic preface. When this was reissued in 1680, he went even further by adding an engraved portrait of Boyle, a re-engraving of the portrait of Boyle produced by William Faithorne in 1664 which is now well known but which had never been published, and which de Tournes really had no authority to use (Plate 1).[34]

Boyle was furious, and he arranged for various public complaints about the de Tournes edition to be issued, starting with a lengthy one written by Oldenburg and published in *Philosophical Transactions* in December 1676: the edition can only just have come to Boyle's attention at that point, thus illustrating the strength of his feelings on the matter. One of his complaints was a legitimate one, namely that de Tournes had overlooked the Latin edition of one of Boyle's key works, *The Origin of Forms and Qualities*. The others, however, were stranger, if characteristic of Boyle's view of his output. Thus he objected that de Tournes had failed to present his works in the order in which they were published, which was true: de Tournes arranged them thematically, starting with pneumatics and hydrostatics, then presenting more programmatic works prior to those dealing with the 'qualities' of things. He was probably trying to make them into a kind of system, as was the case with various posthumous collections of Boyle's writings, in contrast to his own essentially historical view of the way in which his ideas had evolved, which made him wary of collected editions in any case.[35] Boyle also objected that de Tournes failed to state that his books were originally written in English, thus explaining any defects in their Latin style, while he wanted the original publication date of his books to be given so that 'by comparing the several true *Dates* of the first Edition of this Authors works

---

[34]   Ibid., vol. 1, pp. lxxviii–lxxxi. For the portrait, see R.E.W. Maddison, 'The Portraiture of the Hon. Robert Boyle', *Annals of Science*, 15 (1959), 141–214, on p. 156.

[35]   *Works*, vol. 1, pp. lxxx–lxxxi, lxxxv–lxxxvi.

with the Books of others, since printed, the priority of the Experiments, and Considerations, respectively contained in them, may be truly stated'.[36]

Similar objections were made in the prefaces to Latin editions of Boyle's writings over the next few years, in the *Second Continuation* to *Spring of the Air* in 1680, in the *Aerial Noctiluca*, his exploration of phosphorescent substances, published in 1682, and in *Of the Reconcileableness of Specifick Medicines to the Corpuscular Philosophy* in 1686.[37] Boyle's protests succeeded to the extent that de Tournes got the message about the omission of *Forms and Qualities* (which was specifically reiterated in the 1686 version of the comments), since this work was added to his collection in 1687. Otherwise, however, Boyle's reiterated complaints fell on deaf ears, not least since de Tournes's collection did serve a genuine need in the absence of any other collected edition of Boyle: indeed, it was advertised by booksellers even in England, and copies of it survive in various extant English libraries of the period.[38]

But this use of prefaces for polemical purposes on Boyle's part did not originate with the de Tournes episode. Earlier in the 1670s, Boyle had used the prefatory material to various of his books to complain about what he saw as unacknowledged use of his work by others. A classic case was the preface by the 'publisher' to *Mechanical Qualities* of 1675, which objected to the plagiarism of material from Boyle's *Experiments and Considerations touching Colours* (1664) by the miscellaneous writer William Salmon. Salmon evidently felt that, once such material was publicly available in print, it was appropriate for him to use it with only the most generalised of acknowledgments, but Boyle was adamant that such practices represented a 'Usurpation of the Labours of the Benefactors to Philosophy' – a powerful assertion of intellectual property rights.[39] Boyle's 'publisher' made a similar claim against an even more surprising opponent, the French savant Jean-Baptiste du Hamel, in 1680, in this instance mainly concerning unacknowledged use of findings from *The Sceptical Chymist*, which was then just being reprinted nearly twenty years after it had first appeared: Boyle was concerned that 'that if the Reader were not advertis'd, he might easily suspect, that *Mr Boyle* had not lent to, but borrowed of an Author, who appears so capable of enriching the Curious with excellent things of his own'.[40]

As cases like this indicate, Boyle seems to have been concerned that people should be aware both of what his corpus comprised and of when its components had actually been published, and to this end he indulged in an unusual practice in many of his books, from his *New Experiments and Observations Touching*

---

36 Ibid., vol. 1, p. lxxix.
37 Ibid., vol. 9, pp. 128–31, 359–60; vol. 10, pp. xlviii–xlix.
38 Ibid., vol. 1, p. lxxxiv.
39 Ibid., vol. 8, pp. 317–18.
40 Ibid., vol. 9, pp. 29–30.

Cold (1665) onwards, namely of including a list of his writings in chronological order, often accompanied by an apologetic note. The list in *Cold* even gave the exact month in which each book had come out, though most later lists contented themselves with the year of publication. An updated version of the *Cold* list appeared in the Latin translation of Boyle's *Hydrostatical Paradoxes* in 1669 (Plates 5–6) (including manuscript emendations evidently made by the publisher, which are found in most – though not all – extant copies), while a newly compiled list appeared in 1673, in this case in Boyle's *Essays of the ... Nature of Effluviums*. Subsequently, in the aftermath of the de Tournes debacle, such lists proliferated, reaching a climax with the separately published one of 1688–89 which has already been referred to, which was then reprinted in a number of Boyle's books in his last few years.[41]

These complaints and accompanying listings formed part of a somewhat querulous attempt on Boyle's part to assert his originality and to rebut any potential charge that he had plagiarised his ideas from others, which seems to have been a matter of great concern to him. Indeed, quite apart from retrospective complaints like those about Salmon and Du Hamel already cited, another frequent topic in Boyle's or his publishers' prefaces related to overlapping work by other authors, and to the assertion that Boyle had executed his own investigations long before he heard about theirs, even if he had been overtaken in publishing his findings. Such assertions appear in a number of his books, and in at least some cases, including his *Essay about the Origin and Virtues of Gems* (1672) and his *Short Memoirs for the Natural Experimental History of Mineral Waters* (1685), there is reason to believe that one of the factors that impelled Boyle into print on the subject at the point when this occurred was the fact that the publication of related work by others meant that he saw a danger of his own findings being superseded – though, ironically, Boyle need not have worried, since, just as *Gems* offered a highly original analysis of the structure of minerals, so the findings laid out in *Mineral Waters* offered an exemplary programme for the analysis of water from medicinal springs.[42]

Indeed, what is striking about these and other prefaces is the extent to which they display a real sense of insecurity on Boyle's part. Quite apart from this concern to assert his independence from other authors, they also apologise for shortcomings in virtually every aspect of Boyle's books – their prematurity, their incompleteness or their stylistic inadequacy.[43] Indeed, these apologies are all the more striking when set end to end, something which is easy with a

---

[41] Ibid., vol. 1, pp. lxxxi–lxxxiii. For the *Cold* list, see ibid., vol. 4, p. 517; for the *Hydrostatical Paradoxes* list, see ibid., vol. 5, p. xxi; for the *Effluviums* list, see ibid., vol. 7, pp. xxviii, xxx.

[42] Ibid., vol. 7, pp. xi–xii, 5–6 and 3ff., passim; vol. 10, pp. xxix, 207 and 205ff., passim.

[43] See *Scrupulosity and Science*, ch. 7.

uniform modern edition of Boyle's *Works*, but which would have been difficult for Boyle himself, who (as he admitted) did not always have copies of relevant earlier books to hand when he consigned a new one to the press.[44] Some of the problems Boyle outlined may have been well founded, though he does seem to have been strangely accident-prone: the number of his manuscripts which were lost or misordered between leaving him and reaching the printer is extraordinary, as in one instance when one section of a book 'was by the Negligence of him, that should have carried it to the Press, severed from the rest of that Tract, and not seasonably deliver'd to the Printer', meaning that it had to be included in a different volume three years later.[45] In terms of delays to the completion of works, Boyle blamed 'divers Removes, Indispositions of Body, Law-Suits, and other Avocations', while other mishaps included fires adjacent to his lodgings and the suspected theft of his manuscripts by his assistants.[46] One case of the latter was the more bizarre in that, having initially suspected theft when he failed to find a manuscript where he looked for it, Boyle's deduction on subsequently coming across it was that the penitent thief must have returned it.[47]

Here, I think we gain an insight into the slightly troubled personality which stimulated Boyle to achieve the things he did, probably initially stemming from a sense of inadequacy induced by the high expectations of his domineering father, the Great Earl of Cork. This was then exacerbated by Boyle's encounter with the tradition of casuistry considered in Chapter 5, which seems to have made him almost permanently prone to qualification, convolution and self-doubt, and which in turn encouraged him to constant examination both of his conscience and of the world around him.[48] The result is to remind us of a rather different Boyle from the celebrated author of the extensive corpus we have been examining here. But, arguably, the two have to be seen together in order properly to understand him. Boyle was in a sense *driven* to produce the output he did, and he worried incessantly about it even then. This makes it all the more revealing that it is in the prefaces and public statements with which he and his 'publishers' interspersed his major scientific findings that we find hints of the telling kind I have here divulged. Yet this was combined with, and inseparable from, the more strategic uses of print that I outlined earlier in this chapter, and the two need to be seen together if we are ever to do full justice to Boyle and his achievements.

---

44 For example, *Works*, vol. 6, p. 33; vol. 10, p. 356.
45 Ibid., vol. 7, p. 414.
46 Ibid., vol. 9, p. 273; *Scrupulosity and Science*, pp. 135–7, 143–4.
47 *Works*, vol. 10, pp. 305–6.
48 See also *Scrupulosity and Science*, passim.

Chapter 8
Boyle and the Supernatural[1]

On 10 January 1690, Boyle received a visit at the home he shared with his sister, Katherine Jones, Lady Ranelagh, in London's Pall Mall, from the explorer and colonialist Pierre-Esprit Radisson. As in many similar cases, he recorded the conversation they had in one of the compilations I have called his workdiaries, which have already been more than once referred to in this volume: these were sheaths of notes in which Boyle recorded experiments or observations he made, or reports he received from others, and there is now a complete online edition of them.[2] What struck me in this instance was Boyle's note that 'among other discourse that past between us about his journeys farre up into the continent of the northern America I desird him when our Company withdrew to tell me freely whether he had observd any thing of supernaturall among the savages he had conversd with.'[3]

I will return to the content of the conversation that followed, and also to the significance of Boyle's waiting till the company withdrew to put this question to Radisson, but what is striking is the directness with which Boyle solicited information on phenomena that he explicitly described as supernatural. This is paralleled in other instances in workdiaries of Boyle's later years. For instance, under the date 15 February 1689 he recorded a similar encounter with another colonialist, Richard Cony:

> Yesterday I questioned Colonel Cony, recently returned from governing the islands of Bermuda, as to whether among the African peoples, with whom he has had more contact than other Europeans, he had observed anything supernatural, seeing that they are held to be so skilled in the magic art'

And under 18 January 1690 we read:

> Yesterday in the afternoon I receivd a visit from Mr [here there is a blank in the manuscript] a Captain of & nominated to be a lieutenant Colonell of protestant

---

[1] Previously published in a French translation by Charles Ramond as 'Boyle et le surnaturel', in Myriam Dennehy and Charles Ramond (eds), *La Philosophie Naturelle de Robert Boyle* (Paris: Librairie Philosophique J. Vrin, 2009), pp. 213–36.
[2] See http://www.livesandletters.ac.uk/wd/index.html, and *Boyle Papers*, ch. 3.
[3] Workdiary 36–102.

Swisses who appearing to me upon discourse to be a man of parts & serious I was thereby invited to aske him what I might beleeve of some relations that I heard he had made to some of his freinds of my acquaintance concerning supernaturall occurrences.[4]

I confess that I was surprised by the bluntness and clarity with which these requests were framed, which is echoed in other scattered documents from Boyle's later years,[5] and it set me thinking about the meaning and significance of such statements on Boyle's part. What did he mean by 'supernatural' in this context, and how did it relate to the way the term was employed both elsewhere in Boyle's own writings and also in contemporary usage – since the implication is that, in using this term in conversation with his informants, he presumed that it would be obvious to them what he meant by it. Why was he so keen to gather eye-witness reports of such phenomena? And how did this fit into his overall outlook?

First, therefore, let me offer some general considerations about Boyle's use of the term 'supernatural' and the way in which this fitted into his philosophy as a whole. Here I, like other authors on Boyle, will put together a montage of quotations from sources dating from various points in Boyle's life – which has its pitfalls, but which is the obvious default position for understanding his ideas on the topic in question unless there is clear evidence that these changed in the course of his career. It is perhaps worth noting that 'An Essay Entitul'd about things naturall & some supernatural', which is listed in one of Boyle's inventories of his unpublished writings in his later years, unfortunately no longer survives.[6]

---

4   Extract from Latin translation of a lost workdiary printed in *Scrupulosity and Science*, pp. 245, 248; Workdiary 36–103 (the figure involved in this case was perhaps Sieur Sigismund Derlack, who was given a pass on 17 June 1690 to recruit soldiers in the Protestant cantons; the pass notes that he had served as a captain in Colonel John Beaumont's regiment: *Calendar of State Papers Domestic 1690–91*, p. 35).

5   See, for example, Royal Society MS 191, fol. 92: 'A Dr of Physick ingenious learned, sober, reserved & indeed one of the most accomplisht persons I ever met with of his laudable profession, having visited this countrey among divers which his curiositie leads him to travell into, did me the favour to visit me more then once during his stay in London (which I wisht had been longer). And one tyme discoursing freely with him of some uncommon subjects, I was a little surprised to hear such a person as he, (who is both conscientious & judicious) declare to me that he had not long before at a place I was no stranger to & in the company of divers virtuosi, seen some experiments or phænomena supernaturall.' At this point, the text breaks off. Within it, 'indeed' is followed by 'a most accomplisht pe' deleted; a letter is deleted before 'ever' and another before 'And'; 'in the' is deleted after 'before'; and 'seen' is altered from 'seeing'.

6   *Works*, vol. 14, p. xlvi.

Most obviously, the term 'supernatural' was one that Boyle used in relation to God and his attributes, and to the revelation from God vouchsafed us in the scriptures. Thus, in an until recently unpublished text on the compatibility of reason and religion, Boyle writes of 'the supernatural truths that we Christians believe are revealed in the Scriptures', while at one point in the first part of *The Christian Virtuoso* (1690–91), he itemised 'the things that Reveal'd Religion declares, (such as are the Decrees, the Purposes, the Promises, *&c.* of God, and his most peculiar manner of Existing and Operating)' – 'things so Sublime and Abstruse, that they may well be look'd upon as of an higher Order than merely Physical Ones, and cannot be Satisfactorily reach'd by the mere Light of Nature'. In another significant passage in the same work, he divided 'Experience' up into '*Personal, Historical*, and *Supernatural*, (which may be also styl'd *Theological*)'.[7]

This meant that there was a whole range of phenomena which Boyle considered important but which were beyond the cognisance of natural philosophy. Apart from scriptural revelation, there was also the entire economy of angels and demons by whom he thought God's world was populated. In his *Of the High Veneration Man's Intellect owes to God; Peculiarly for his Wisedom and Power* (1684–85), Boyle explained how:

> we may rationally suppose, that if we were quick-sighted enough to discern the Methods of the Divine Wisedom in the Government of the Angelical and of the Diabolical Worlds, or great Communities, if I may so call them; we should be ravish'd into admiration how such Intelligent, Free, Powerfull, and Immortal Agents, should be without violence offer'd to their Nature, made in various manners to conspire to fulfill the Laws, or at least accomplish the Ends, of that great *Theocracy*, that does not alone reach to all kinds of bodies, to Men, and to this or that rank of Spirits, but comprises the whole Creation, or the great Aggregate of all the Creatures of God.

Indeed, he went so far as to argue that:

> the government of one Dæmon, may be as difficult a work, and consequently may as much declare the Wisedom and Power of God, as the government of a whole *Species* of inanimate bodies, such as Stones or Metals; whose Nature determines them to a strict conformity to those primordial Laws of Motion, which were once settled by the great Creatour, and from which, they have no Wills of their own to make them swerve.[8]

---

7   Ibid., vol. 14, p. 216; vol. 11, p. 324, 307. Cf. ibid., vol. 11, p. 303.
8   Ibid., vol. 10, pp. 177–8.

In general, of course, these were topics which he wrote about only in writings defined as being 'Theological', which were carefully differentiated from his 'Philosophical Books and Tracts' in the lists of these circulated in his later years.[9] Elsewhere, as in *The Origin of Forms and Qualities* (1666–67), he explained how his 'Business, in this Tract, is to discourse of Natural Things as a Naturalist, without invading the Province of Divines, by intermedling with Supernatural Mysteries'.[10] But even this distinction was not an entirely watertight one, since he believed that there were certain truths relating to nature which were only available from such sources. 'We may farther consider', he wrote in *The Excellency of Theology, compar'd with Natural Philosophy* (1674), 'That as to things Corporeal themselves, which the Naturalist challenges as his peculiar Theme, we may name particulars, and those of the most comprehensive nature, and greatest Importance, whose knowledge the Naturalist must owe to Theology.' These included such facts as God's creation of the world, its relative novelty and its ultimate ending, and elsewhere he was implicitly critical of 'the *Cartesian* Philosophers, who lay aside all Supernatural Revelation in their Inquiries into Natural things'.[11]

'Supernatural' is also a term that Boyle repeatedly used in relation to miracles: indeed, a word search for the term in the conveniently available InteLex electronic version of his *Works* turns it up in this context more than almost any other; he also repeatedly used it in this connection in the hitherto unpublished writings on related subjects of which Jack MacIntosh has published a comprehensive selection in his *Boyle on Atheism*. In fact at one point Boyle referred to 'Miracles or Phænomena supernatural' as if the two were synonymous.[12] As he wrote in one of his manuscripts on miracles:

> Nor are they only more sublime than the Objects of Naturall Philosophy, but of greater Importance to us, since the knowledg & use of Corporeal things dos chiefly relate to our Bodies, & usually reaches but to this Life. Whereas to have a right Judgement of Miracles & their consequences is of very great moment, if not necessity, to direct us securely in making our Choyce of Religion which is the importantst action of our Understanding, & on which very much depends, besides the solid Happiness of the Soul in this Life, the endless felicity of the whole man in the Life to come.[13]

---

9   Ibid., vol. 11, pp. 187ff.
10   Ibid., vol. 5, p. 293.
11   Ibid., vol. 8, pp. 20–22, 258.
12   J.J. MacIntosh (ed.), *Boyle on Atheism* (Toronto, 2005), p. 261.
13   Ibid., p. 274.

In his view, 'True Miracles are the proper & greatest Proofes, that the Message or Doctrine that is attested by them, is indeed a Divine Revelation'.[14] The most significant miracles were clearly those associated with Christ himself, and perhaps particularly the miracle of the Pentecost, about which Boyle wrote at length. But he also included as '*supernatural* Things' 'true Prophecies of Unlikely Events, fulfill'd by Unlikely Means', such as the extraordinary triumph of Christianity over the centuries, which had the advantage over '*transient*' miracles (like turning water into wine) that they were '*permanent*'.[15]

These were events which overruled 'the Ordinary Course of Nature', but not 'the Essential Nature of Things', and Boyle was anxious not to make unnecessary claims for the extent of such 'supernatural' interventions. He argued that God was economical in the extent to which he overrode nature's normal course, 'forbearing to display his Almighty power, save in those parts of it, that must necessarily be miraculous, because they require a power surpassing that of ordinary nature'.[16] Indeed, since the whole point of his preoccupation with miracles was the evidence they provided of God's power in distinction from the 'natural' causes according to which the universe normally operated, it is important to stress that Boyle seems to have had a fairly clear, if not straightforward, sense of a distinction between the natural and the supernatural. To take an example almost at random, in the definitional section on '*Nature*' in *A Free Inquiry into the Vulgarly Receiv'd Notion of Nature* (1686), he distinguished between 'Cures wrought by Medicines, [which] are *Natural* Operations' and 'the miraculous ones, wrought by Christ and his Apostles, [which] were Supernatural'.[17] Matters are complicated by a passage in the second part of *The Christian Virtuoso* published in the 1744 edition of Boyle, in which, 'instead of dividing the operations of God, here below, into two sorts only, natural and supernatural', he suggested that 'we may take in a third sort, and divide the same operations into supernatural, natural in a stricter sense, that is mechanical, and natural in a larger sense, which I call supra-mechanical'.[18] However, as will be seen, this simply extended the boundaries of the natural, and we should not be under any illusions about the sheer complexity which Boyle thought was likely to characterise the natural realm, including a range of intermediate powers and principles which went beyond the purely mechanical, such as the weight and elasticity of the air, or the role of fermentation or gravity.[19]

---

[14] Ibid., p. 275.

[15] *Works.*, vol. 11, pp. 321–2. For Boyle's view on miracles, see above, p. 121, and the references there cited.

[16] MacIntosh, *Boyle on Atheism*, pp. 262–3, 266.

[17] *Works*, vol. 10, p. 453.

[18] Ibid., vol. 12, p. 477.

[19] See Antonio Clericuzio, 'A Redefinition of Boyle's Chemistry and Corpuscular Philosophy', *Annals of Science*, 47 (1990), 561–89; Peter Anstey, 'Robert Boyle and the

On the other hand, it is worth making it clear here that, although Boyle seems to have been sure, concerning the primary principles of matter and motion, that motion was due to 'some Immaterial Supernatural Agent' rather than innate in matter itself, I am unhappy about the emphasis on 'radically supernaturalistic' elements in his thought in the interpretation of Keith Hutchison.[20] Thus, in *The Origin of Forms and Qualities*, Boyle distanced himself from hypotheses which would 'put Omnipotence upon working I know not how many thousand Miracles every hour, to perform that (I mean the Generation of Bodies of new Denominations) in a supernatural way, which seems the most familiar effect of Nature in her ordinary course'.[21] Granted that Boyle saw it as a given that the laws of nature themselves depended on God's power – so that, theoretically speaking, his view of the world was almost totally supernaturalist – in practice, his separation of the 'natural' from the 'supernatural' means that we can follow him in using this as a worthwhile diagnostic tool.

Indeed, in various passages in his natural philosophical writings, Boyle was at pains to minimise the intervention of the 'supernatural' in the operations of the world. Take, for instance, what he says in *Experiments and Considerations touching Colours* (1664) concerning the 'Opinion concerning the Complexion of *Negroes*, that is not only embrac'd by many of the more Vulgar Writers, but likewise by that ingenious Traveller Mr. *Sandys* [who] would have the Blackness of *Negroes* an effect of *Noah*'s Curse ratify'd by God's, upon *Cham*'. Boyle's view was that:

> though I think that even a Naturalist may without disparagement believe all the Miracles attested by the Holy Scriptures, yet in this case to flye to a Supernatural Cause, will, I fear, look like Shifting off the Difficulty, instead of Resolving it; for we enquire not the First and Universal, but the Proper, Immediate, and Physical Cause of the Jetty Colour of *Negroes*.

He then went on to elucidate this in terms of 'some Peculiar and Seminal Impression', adding a range of acute observations about the actual nature of pigments in the skin.[22]

Heuristic Value of Mechanism', *Studies in History and Philosophy of Science*, 33 (2002), 161–74.

[20]   *Works*, vol. 8, p. 105; Keith Hutchison, 'Supernaturalism and the Mechanical Philosophy', *History of Science*, 21 (1983), 297–333. See also MacIntosh, 'Locke and Boyle', pp. 206–7; Peter Anstey, *The Philosophy of Robert Boyle* (London, 2000), esp. pp. 163–5, 207–8.

[21]   *Works*, vol. 5, p. 342.

[22]   Ibid., vol. 4, pp. 88, 89 and 88–93, passim. See also Cristina Malcolmson, *Studies of Skin Color in the Early Royal Society: Boyle, Cavendish, Swift* (Farnham, 2013), pp. 31ff. and passim.

He took a similar line – though complicated by typical Boylean ambivalence – in what he said in his 'Experimental Discourse of Some Unheeded Causes of the Insalubrity and Salubrity of the Air' (1685) about 'the Controversie about the Origine of the Plague, namely, Whether, it be Natural, or Supernatural', arguing that:

> neither of the contending Parties is altogether in the right: since 'tis very possible, that some Pestilences may not break forth, without an extraordinary, though perhaps not Immediate interposition of Almighty God, provok'd by the Sins of Men: and yet other Plagues may be produc'd by a Tragical concourse of merely Natural Causes.

On balance, Boyle was critical of authors like Isbrandus de Diemerbroeck in his *De peste libri quatuor* (1644), 'who derives the Plague from a *Supernatural Cause, the wrath of God against the sins of men*', on the grounds that:

> it seems unphilosophical, and perhaps rather seems than is very pious, to recur without an absolute necessity to Supernatural Causes, for such Effects as do not manifestly exceed the power of Natural ones: though the particular manner of their being produc'd, is perchance more than we are yet able clearly to explicate.

He added:

> I think it the more questionable, whether all Plagues are Supernatural Exertions of God's Power and Wrath against the Wicked, because I observe that *Brutes* (which are as well uncapable of moral Vice, as moral Vertue) are yet oftentimes subject to *Murrains*.[23]

I would also like to draw attention to two episodes evidenced by Boyle's correspondence when potential 'supernatural' explanations for phenomena were canvassed, and when he seems to have been reluctant to accept them. The first was the notorious episode involving Valentine Greatrakes, the Irish 'stroker' who came to England and exercised his healing powers in 1666.[24]

---

[23]　Ibid., vol. 10, pp. 329–30, 332. Cf. I. De Diemerbroeck, *De peste libri quatuor* (Arnhem, 1646), esp. pp. 18ff.

[24]　See Michael McKeon, *Politics and Poetry in Restoration England* (Cambridge, MA, 1975), pp. 208–15; J.R. Jacob, *Robert Boyle and the English Revolution* (New York, 1977), pp. 164–76; J.R. Jacob, *Henry Stubbe, Radical Protestantism and the Early Enlightenment* (Cambridge, 1983), pp. 50–63, 164–74; A.B. Laver, 'Miracles No Wonder! The Mesmeric Phenomena and Organic Cures of Valentine Greatrakes', *Journal of the History of Medicine*, 33 (1978), 35–46; Eamon Duffy, 'Valentine Greatrakes, the Irish Stroker: Miracle, Science and Orthodoxy in Restoration England', *Studies in Church History*, 17 (1981), 251–73;

Greatrakes himself was convinced, of his ability to heal, that 'there is something in it of an extraordinary Gift of God', and in recounting his cures he constantly invokes 'God's mercy' and 'God's blessinge', assuring Boyle that 'to speake the truth to you I have found the power of God wonderfully assisting of me'.[25] Some contemporary commentators similarly considered the cures implicitly comparable with the miracles of Christ and his apostles, like one who assured his readers concerning Greatrakes: 'It is clear, that these 1500 years at least, we cannot hear of any instance of the like nature.'[26] Others took a different view, including Henry Stubbe in his published pamphlet on the affair, *The Miraculous Conformist* (1666), in which, though conflating Greatrakes's powers with those of Christ and the apostles, he argued that they were natural.[27] This pamphlet was addressed to Boyle, who therefore penned a lengthy response which survives in two copies that he retained, though whether he ever sent it is unclear. In it, Boyle was at pains to differentiate Greatrakes's healing powers from those of the apostles, declaring himself 'not yet fully convinc'd that there is ... any thing that is purely supernaturall' about the episode. Though he added a significant rider that 'in the way wherein he was made to take notice of his Guift, & exercise it there may be something of that kind', his verdict was: 'and therefore till the contrary doth appear, I hold it not unlawfull to endeavour to give a Phisicall Account of his Cures', adducing an explanation in terms of Greatrakes's touch being 'a more noble Specifick'.[28]

N.H. Steneck, 'Greatrakes the Stroker: The Interpretations of Historians', *Isis*, 73 (1982), 161–77; B.B. Kaplan, 'Greatrakes the Stroker: The Interpretations of His Contemporaries', *Isis*, 73 (1982), 178–85; D.P. Walker, 'Valentine Greatrakes, the Irish Stroker and the Question of Miracles', in *Mélanges sur la Littérature de la Renaissance à la Mémoire de V.-L. Saulnier* (Geneva, 1984), pp. 343–56; Simon Schaffer, 'Regeneration: The Body of Natural Philosophers in Restoration England', in Christopher Lawrence and Steven Shapin (eds), *Science Incarnate: Historical Embodiments of Natural Knowledge* (Chicago, IL, 1998), pp. 83–120, on pp. 105–16; C.S. Breathnach, 'Robert Boyle's Approach to the Ministrations of Valentine Greatrakes', *History of Psychiatry*, 10 (1999), 87–109; Jane Shaw, *Miracles in Enlightenment England* (New Haven, CT and London, 2006), ch. 4; and Peter Elmer, *The Miraculous Conformist: Valentine Greatrakes, the Body Politic, and the Politics of Healing in Restoration Britain* (Oxford, 2013).

[25] *A Brief Account of Mr Valentine Greatrak's, And Divers of the Strange Cures By Him Lately Performed* (London, 1666), p. 34 and passim; *Correspondence*, vol. 4, pp. 98–103. For Boyle's own notes on Greatrakes's cures, see Workdiary 26, and above, p. 87.

[26] [Lionel Beacher], *Wonders if not Miracles or A Relation of the Wonderful Performances of Valentine Gertrux [sic]* (London, 1665), p. 5.

[27] Henry Stubbe, *The Miraculous Conformist: or An Account of severall Marvailous Cures Performed by the Stroaking of the Hands of Mr Valentine Greatarick* (Oxford, 1666), pp. 26–7 and passim.

[28] *Correspondence*, vol. 3, pp. 93–107.

The second episode arose in the course of the correspondence Boyle conducted with prominent figures in New England, including the missionary John Eliot and the Governor of Connecticut John Winthrop, who clearly valued his links with Boyle not least for the expert advice Boyle could provide on matters of natural philosophy. In autumn 1670, both Eliot and Winthrop wrote to Boyle about a strange occurrence at a pond at Watertown, Massachusetts, where (to quote the account Boyle had transcribed into one of his workdiaries) 'on a sudden all the fish dyed and thrust themselves out of the water to dye on the shore ... the cattle refused to drinke of the water for 3 dayes, but after 3 dayes dranke of it again'.[29] The fact that this is one of the very few of Eliot's extant letters to Boyle that mentions such a phenomenon evidently indicates its impact, and Eliot had no doubt that it was a supernatural event, opening his account of it by saying: 'There hath bene a rare work of God this Summer.'[30] Winthrop, on the other hand, was more ambivalent. To some extent he seems to have been looking for a natural, comparative explanation, perhaps to undercut the supernaturalist explanations put forward by the Indians, who averred the unprecedented nature of the event and saw it as 'a strange & prodigious thing'. But, explaining how 'there be many difficulties, in the consideration about this, as to any naturall cause of it', in his letter to Boyle he was therefore:

> bold humbly to crave your Judicious thoughts of this thing, whether any naturall cause can be knowne, and whether any memory of this age, or history of former of the like in any lakes in Ireland, or ponds or lakes in other parts of the world hath observed apparent causes, or whether only to be looked upon as supernaturall.[31]

Unfortunately, we lack Boyle's response to Winthrop's letter, but his interest in the phenomena involved is clear not only from the record in his workdiaries, but also from a separate commentary either on this or a very similar episode, based on 'a more circumstantial account of this Affaire' that he received from another New Englander who visited him. Moreover, the context of this *is* the kind of naturalism that Winthrop evidently expected from Boyle. Noting how the discolouration of the lake and the poisoning of the fish occurred 'In an Autumn that followed a very dry season', Boyle deduced that 'this change was produc'd by the unwonted invasion of subterraneal *Effluviums*'.[32] As far as

---

[29]   Workdiary 21–402. Boyle's interest in this is shown by the fact that this workdiary also includes a later fair copy in the hand of another amanuensis, entry 402b.

[30]   *Correspondence*, vol. 4, p. 188.

[31]   Ibid., pp. 183–7: there are in fact two overlapping versions of this letter, both of which are published in *Correspondence*. It is perhaps worth noting that the Indians suggested a link with lightning, which Winthrop ignored.

[32]   Royal Society MS 199, fol. 144, copied in BP 21, fol. 113, and printed from there in *Works*, vol. 13, p. 412.

Winthrop is concerned, a further, similar episode arose at much the same time when he reported in a letter to the virtuoso Lord Brereton which was read out at the Royal Society 'a very strange and prodigious wonder', in the form of a piece of ground at Kennebunk, Maine which apparently moved: in this case, he seemed tempted to a supernaturalist explanation, but was rather brought down to earth by the matter-of-fact response of Henry Oldenburg.[33]

There can be no doubt that the predominant trend in late seventeenth-century England, both in the general culture and among natural philosophers, does seem to have been towards a preference for natural as against supernatural explanations. To illustrate this, one need only instance books like Thomas Sprat's *History of the Royal Society* (1667) or Edward Stillingfleet's *Origines Sacrae* (1662 and subsequent editions), or the controversies that occurred at this time on such topics as the geological history of the world, the status of miracles or the nature of mental illness.[34] In Boyle's case, an example – on which I have commented elsewhere – is provided by the explanation given by such modish Latitudinarian divines as Stillingfleet and Gilbert Burnet of his bouts of religious melancholy, which Boyle clearly believed represented direct 'Injections' by the Devil. For they put them down to 'a Feaver or a Vapour of the Spleen', or 'depressions or weaknesses of the Animal Spirits oftentimes proceeding from the want of Nourishment or Free Air or Exercise or pleaseing Circumstances &c.'[35]

Boyle, therefore, was ambivalent. However open he may have been to the idea that there might be a wide range of subtle explanations within nature, like the effluvia he invoked to account for the discoloured lake in Massachusetts, the general naturalising trend was one about which he had reservations. This is seen

---

[33] See Birch, *Royal Society*, vol. 2, pp. 473–4 (and Royal Society Early Letters W.3.24); A.R. and M.B. Hall (eds), *The Correspondence of Henry Oldenburg* (13 vols, Madison, WI and London, 1965–86), vol. 7, p. 569. For a more naturalistic explanation, see Increase Mather, *An Essay for the Recording of Illustrious Providences* (1684), reprinted as *Remarkable Providences Illustrative of the Earlier Days of American Colonisation*, ed. George Offer (London, 1890), pp. 228–9; but for a 'supernaturalist' response to Winthrop on the part of Fellow of the Royal Society Theodore Haak, see R.C. Winthrop (ed.), *Correspondence of Hartlib, Haak, Oldenburg and others of the Founders of the Royal Society, with Governor [John] Winthrop of Connecticut 1661–72* (Boston, MA, 1878), pp. 45–6.

[34] For miracles, see above, pp. 121, 166–7, and Shaw, *Miracles in Enlightenment England*. For geological controversy, see, for example, Roy Porter, *The Making of Geology: Earth Science in Britain 1660–1815* (Cambridge, 1977), and Paolo Rossi, *The Dark Abyss of Time: The History of the Earth and the History of the Nations from Hooke to Vico* (English translation, Chicago, IL, 1985). For mental illness, see Michael MacDonald, 'Religion, Social Change and Psychological Healing in England, 1600–1800', *Studies in Church History*, 19 (1982), 101–25. In general, see R.E. Sullivan, *John Toland and the Deist Controversy* (Cambridge, MA, 1983).

[35] *Scrupulosity and Science*, pp. 83–5, 89, 90-1, and ch. 4, passim.

in his letter to Stubbe about the Greatrakes episode, in which he complained about such tendencies, particularly in the extreme form these reached among contemporary free-thinkers, deploring the fact that he lived:

> in an Age, where so many doe take upon them to deride all that is supernaturall; &, whilst they loudly cry up Reason, make no better use of it, then to imploy it, first to Depose Faith, and then to serve their Passions & Interests.[36]

Boyle's acute awareness of such trends in the thought of the day, and his anxiety to delimit them, is clearly seen in a public debate which occurred in England at this time about the reality and nature of another kind of ostensibly supernatural phenomenon, namely witchcraft. Boyle's views are evidenced by the letters he exchanged with the divine Joseph Glanvill in 1677–78 in connection with Glanvill's collection of 'relations' for his demonological treatise, of which an extended version was posthumously published in 1681 as *Saducismus Triumphatus*. As Boyle put it: 'any one relation of a supernatural phænomenon being fully proved, and duly verified, suffices to evince the thing contended for; and, consequently, to invalidate some of the atheists plausiblest arguments'.[37] Interestingly, in this connection he also invoked alchemy, in the form of the supposed transmutation of base metal into gold recently carried out by Wenzel Seiler in the presence of the Holy Roman Emperor, and Lawrence Principe has suggested that it was partly because alchemy offered an interface between the natural and the supernatural that Boyle was so interested in it. This is highly plausible, although in practice Boyle failed to deploy it for overtly apologetic purposes, either because he never actually achieved the results he was hoping for, or because of his moral scruples about such intercourse with the supernatural to which I have drawn attention elsewhere.[38]

Whatever the moral implications, Boyle clearly *did* believe that it was possible for humans to have contact with supernatural agents in the form of angels or demons. He was convinced that men might 'aspire to the knowledge & conversation of incorporeal spirits; & <of> that angelical Community (if I may so call it) that consists of Rational & Immortal beings not clog'd with visible Bodys'.[39] Lawrence Principe has published in full a dialogue on the legitimacy or

---

[36] *Correspondence*, vol. 3, p. 95.

[37] Ibid., vol. 4, pp. 456–7.

[38] Lawrence M. Principe, *The Aspiring Adept: Robert Boyle and his Alchemical Quest* (Princeton, NJ, 1998), ch. 6 (on Boyle's interest in Seiler, see ibid., pp. 95–7); *Scrupulosity and Science*, ch. 5. Principe's suggestion that Boyle's anxieties might have been resolved by finding a more amenable advisor than Burnet in his later years is unfortunately rendered less likely by the fact that Burnet was still advising Boyle within months of his death (ibid., pp. 79–80, 90–92; Principe, *Aspiring Adept*, p. 205 and n. 95).

[39] MacIntosh, *Boyle on Atheism*, p. 256.

otherwise of intercourse with angels, while a further interesting paper survives in which Boyle answered various arguments against the possibility of such contact, largely on the grounds that spiritual beings could not interact with humans.[40] He dismissed such objections partly because 'we men understand very little of the nature, customes, & government of the Intelligent creatures of the *spirituall world*: and particularly what concernes the Falne Angells or bad Dæmons', and partly because few denied the prime example of such interaction in the form of the interconnection between the human soul and the body. In his view, 'they that have had the lucke to see the apparitions, & supernatural feates of spirites, may justly beleeve their Eyes & other Senses; notwithstanding the unaccountablenes of the manner of such things'. He added elsewhere that:

> To those that say by way of objection that they would give any thing to see a Spiritt, but could never find any that could show it them. It may be answered that a man may be willing to give any money to see a Comett, but could meet with no body that could shew it him. For as those lights appear but very seldome, and upon conjunctures of Circumstances that we can neither foresee nor command[,] so those Angels of darkness appear but rarely to the eye, and the times [and] places of their Apparitions are not before hand knowable to Us.[41]

By analogy with the way in which the supernatural beings with whom humans might have contact could be either good angels or demons, Boyle seems also to have admitted that the bona fide miracles that he saw as crucial for proving the truth of the Christian religion were matched by analogous occurrences which were also potentially supernatural. Sometimes, it is true, he seems to deny this status to such phenomena. At one point, he spoke of 'the Distinction between Divine Miracles strictly so call'd, and other stupendious things that to most men seem to be supernatural', while in his letter to Stubbe, he had reserved the concept of 'True Miracles' for Christianity, 'for I speak not of Naturall Prodigys, Sorcerys, or Impostures', adding that he was 'little convinc'd that I ought to beleive the Suspitious & unlikely reports that goe of the Miracles of the Turks & Heathens'.[42] Elsewhere, he explained more fully how the 'Divine Miracles' of Christianity were:

> <very differing from such as are but> uncommon workes of meere nature *or* surprizing effects of naturall Magick, *or* the impostures or collusions of Crafty

---

[40] Principe, *Aspiring Adept*, pp. 310–17 (with discussion on pp. 191–3); Macintosh, *Boyle on Atheism*, pp. 256–9 (the text, from BP 2, fols 100–101 and 104–5, is noted in Principe, *Aspiring Adept*, pp. 206–7).

[41] MacIntosh, *Boyle on Atheism*, pp. 260–61.

[42] Ibid., p. 264; *Correspondence*, vol. 3, p. 100.

cheats (such as I take most of the pretended miracles of the heathens and especially of their oracles to have been) ... *or* the powerfull application of the sacred name of God to which some of the jews fondly impute the Miracles of Christ, or a sanative Ideosyncrasie or <happy> peculiar temperament.

Yet he acknowledged that a supernatural element might be involved in the 'demoniacall works performd by Magicians or witches by the help of evill Spirits, such as ware done by the Egyptian sorcerers (if they were not only trikes of Legerdemain or illusions)'. Similarly, when he differentiated the fulfilled 'prophesies of the old and new Testament from Astrologicall predictions and other kinds of naturall Divination as alsoe from those other foretellings of future things', he wavered by adding the rider, 'which pretend whether falsly or truely to be supernaturall such as were the responses given it [*sic*] Delphos and by other heathen oracles', thus leaving open the possibility that such phenomena were genuinely above nature.[43]

Of course, with witchcraft and sorcery, as with miracles, Boyle insisted that the benefit of the doubt was to be given to natural causes. Indeed – quite apart from making allowance for fraud – he more than once invoked witchcraft as an example of an inadequate explanation which should be superseded by a naturalistic one. Thus, in 'About the Excellency and Grounds of the Mechanical Hypothesis' (1674), he explained how 'a sober Physician' would be unlikely to be satisfied with the view of diseases 'That 'tis a Witch or the Devil that produces them; and he will never sit down with so short an account, if he can by any means reduce the extravagant Symptoms to any more known and stated Diseases, as *Epilepsies, Convulsions, Hysterical Fits, &c.*' In *The Origin of Forms and Qualities*, he compared the inadequacy of scholastic 'substantial Forms' as an explanation with the invocation of witchcraft: it 'leaves the curious Enquirer as much to seek for the *causes* and *manner* of particular Things, as Men commonly are for the particular causes of the several strang Things perform'd by Witchcraft, though they be told, that tis some Divel that does them all.'[44] Yet here, too, he seems to have admitted that such phenomena could be supernatural, as in *The Reconcileableness of Reason and Religion* (1675), where he (rather convolutedly) wrote:

That the Supernatural things, said to be perform'd by *Witches* and *Evil Spirits*, might, if true, supply us with *Hypotheses* and *Mediums* whereby to constitute and prove Theories, as well as the *Phænomena* of meer nature, seems tacitely indeed,

---

[43] MacIntosh, *Boyle on Atheism*, pp. 281–2. However, having consulted the manuscript source, I consider that the word he gives as 'secre[t]est' (before 'name of God') is in fact 'sacred' followed by a hash sign at the end of the line, and have here emended the text accordingly; I have also capitalised 'Delphos'.

[44] *Works*, vol. 8, p. 108; vol. 5, p. 352. Cf., for instance, vol. 3, p. 422; vol. 7, p. 267.

but yet sufficiently, to be acknowledg'd, by those modern Naturalists, that care not to take any other way to decline the Consequences that may be drawn from such Relations, than sollicitously to shew, that the Relations themselves are all (as I fear most of them are) false, and occasion'd by the Credulity or Imposture of Men.[45]

This therefore brings us back to the interviews in the late workdiaries with which I began this chapter, and I now want to say a little more both about these instances, and about their rationale. The probable intended destination of these accounts was a compilation that Boyle planned and partially executed under the title 'Strange Reports', a collection of authenticated accounts of out-of-the-ordinary phenomena. The first part of this was published in *Experimenta & Observationes Physicæ* (1691), the last of Boyle's books to come out during his lifetime, but the work had been planned at least since the 1660s, and various materials for it survive among Boyle's extant manuscripts.[46] In part, 'Strange Reports' was intended as a compendium of phenomena that were natural but abnormal, reflecting Boyle's invocation of forces beyond the straightforwardly mechanical that has already been outlined. In this, it reflected an interest in the 'preternatural' that is central to English natural philosophy of this period, the rationale being provided by Francis Bacon, whose example Boyle had fulsomely cited in the 'Advertisements' to *Experimenta & Observationes Physicæ* as a whole.[47] In his *New Organon*, Bacon had stressed the importance of studying nature where she 'strays and turns aside from her ordinary course': 'we must put together an accumulation or particular natural history of all the monsters and prodigious births of nature; and then of everything in nature that is novel, rare or unusual', he wrote, on the grounds that thus we might better understand nature's ordinary processes. Indeed, this emphasis on the study of the 'preternatural' has been seen by historians as a crucial part of Bacon's legacy in terms of extending the explanatory competence of natural philosophy.[48]

---

[45]   Ibid., vol. 8, pp. 278–9.

[46]   *Scrupulosity and Science*, pp. 228ff. For the earliest reference to such a work, see the mnemonic list of Boyle's writings of *c.* 1670, of which 'The Twenty fourth strange Narratives affect': *Works*, vol. 14, p. 336.

[47]   *Works*, vol. 11, pp. 373–6.

[48]   Bacon, *New Organon*, ii, 29, in Graham Rees with Maria Wakely (eds), *The Instauratio Magna, Part II: Novum Organum and Associated Texts* (Oxford, 2004), pp. 298–9. See also Stuart Clark, *Thinking with Demons: The Idea of Witchcraft in Early Modern Europe* (Oxford, 1997), esp. pp. 252–5; Lorraine Daston, 'The Factual Sensibility', *Isis*, 79 (1988), 452–67; Lorraine Daston and Katherine Park, *Wonders and the Order of Nature 1150–1750* (New York, 1998), ch. 6; and Lorraine Daston, 'Preternatural Philosophy', in Lorraine Daston (ed.), *Biographies of Scientific Objects* (Chicago, IL, 2000), pp. 15–41 (including a reference on Boyle's 'Strange Reports' on p. 36, n. 95).

In the case of 'Strange Reports', Boyle also – perhaps more surprisingly – cited 'the Example and Authority of *Aristotle*', alluding to the pseudo-Aristotelian treatise *De mirabilibus auscultationibus*, a collection of bizarre and puzzling phenomena ranging from mines that renewed themselves to strange poisons and cures.[49] This compilation from Theophrastus and other sources is not now thought to have any connection with Aristotle, but in Boyle's period, opinion on it was divided, and Boyle himself clearly thought it was authentic.[50] In an unpublished version of his preface, Boyle expanded the rather spare comments in the published text to explain how his aim, following Aristotle, was to show the use that might be made of a collection of strange reports 'to suggest new <& more reaching> notions about severall things, & to free the understanding from being <too much> straitend by a confinement to common things & the phenomena of nature & art familiar to us'.[51] In other words, perhaps slightly surprisingly, Boyle found an Aristotelian mandate for what has usually been attributed to Bacon.

The subject matter of the section of Boyle's 'Strange Reports' that was published in 1691 reflected just such an interest in the abnormal but putatively natural. It comprised 10 stories of peculiar phenomena, for instance plants in glass jars that could be resuscitated by heat, or stains left by plague on the walls of rooms, or a solution which expanded when exposed to moonlight. The manuscript versions of certain of these survive in juxtaposition with other comparable accounts in a workdiary dating from the late 1660s. For instance, one illustrated how human organs could react to extreme conditions by instancing a 'Major' (probably Edward Halsall) 'accus'd at Madrid of a State crime', who was locked up in a pitch black dungeon for twenty months and found after a while that his eyes became so accustomed to the darkness that he could see well enough to pour out drinks and even catch mice. 'I ask'd the Major how long it was before he began to discern things in the dark,' Boyle notes, continuing:

> Hee answer'd that it was about the 7ᵗʰ moneth of his imprisonment tho he then saw things but very dimly. I ask'd him also whether at last hee could see so well as to judg himself able to read & write if he had bin permitted the use of pens & books? To which hee answer'd he made no doubt he could.[52]

---

49   *Works*, vol. 11, p. 429.

50   See L.D. Dowdall, trans., 'De mirabilibus auscultationibus', in W.D. Ross (ed.), *The Works of Aristotle*, vol. 6, 'Opuscula' (Oxford, 1913). For opinions on its authenticity, see the edition by Johann Beckmann (Göttingen, 1786), esp. p. xvii.

51   *Works*, vol. 11, p. lxviii. Boyle was arguably here invoking Aristotle the natural historian rather than Aristotle the system-builder, a distinction made in *The Origin of Forms and Qualities*: ibid., vol. 5, p. 295.

52   Works, vol. 11, pp. 431–7; Workdiary 24–7 and passim.

In such cases, Boyle's concern (like Bacon's) was with unusual occurrences which seemed likely to elucidate the complex way in which nature operated, but which did not transcend the natural. However, the work was not intended to be limited to the natural, as Boyle made clear in the preface to the published version of 'Strange Reports', where he explained that though only one part was included there, the work was to comprise two components, the first 'relating to things purely Natural, and the other consisting of *Phænomena*, that are, or seem to be, of a Supernatural Kind or Order'.[53] This published statement is worth pausing over: here, Boyle quite explicitly introduces the concept of 'supernatural phenomena' in the context of data-collection. Supreme empiricist that he was, Boyle thought that empirical data could as readily be brought to bear to prove the reality of 'supernatural' events as 'natural' ones: he was therefore keen to collect incontrovertible accounts of strange phenomena which defied any kind of natural explanation as evidence that a genuinely 'supernatural' realm did indeed exist.

Here, we need to clarify the distinction between Boyle's use of 'supernatural' and 'preternatural', since it has been obscured by certain recent writers who have failed properly to differentiate Boyle's usage of the words. Boyle used 'preternatural' to denote phenomena which transgressed the ordinary course of nature but were still perfectly natural. On one occasion in the workdiaries, the word 'preternatural' was deleted and replaced by 'Unusual', and this is typical of his usage, as when in his published writings he referred to 'Persons, who get a *Praeter-natural* Thirst with over-much Drinking', or when 'a Paradox of the Natural and Preternatural State of Bodies' alluded to violent motion.[54] Such commentators as Stuart Clark, Lorraine Daston and Simon Schaffer have focused on Boyle's interest in the 'preternatural', and they have seen him as so fully a party to the increasing naturalisation of such phenomena in his period to which I have already alluded that they have almost ignored Boyle's belief that there was a genuine supernatural realm.[55] In fact, however, Boyle not only considered supernatural phenomena intrinsically significant; he also, particularly in his later years, seems to have come to believe that natural philosophy could serve an important apologetic role by authenticating them, in resistance to a growing

---

[53] *Works*, vol. 11, p. 429.

[54] Workdiary 37–99; *Works*, vol. 10, p. 539; vol. 7, pp. 419ff. For a single instance which I judge ambiguous (out of over 40 in the *Works* as a whole), see ibid., vol. 8, p. 311.

[55] Clark, *Thinking with Demons*, pp. 264, 305–6 and passim; Lorraine Daston, 'Marvelous Facts and Miraculous Evidence in Early Modern Europe', in P.G. Platt (ed.), *Wonders, Marvels and Monsters in Early Modern Culture* (Newark, DE, 1999), pp. 76–104; Simon Schaffer, 'Occultism and Reason', in A.J. Holland (ed.), *Philosophy: its History and Historiography* (Dordrecht, 1985), pp. 117–43; Simon Schaffer, 'Godly Men and Mechanical Philosophers: Souls and Spirits in Restoration Natural Philosophy', *Science in Context*, 1 (1987), 55–85. In fact, Schaffer is the chief source of Clark's views on this topic.

tendency in educated opinion to reject all such phenomena as implausible – a proclivity which he saw as encouraging irreligion and to which he was therefore strenuously opposed. As he put it concerning such sceptics:

> For if these men were satisfyed in their Scruples about the very Notion and Existence of a Spirit or incorporeal Substance Intelligent and powerfull, and that an Immaterial Substance can worke upon matter, I should thinke them more than halfe way advanced towards the acknowledgement of the Being of God.[56]

He elaborated on his motives in the preface he prepared for the section of 'Strange Reports' devoted to supernatural phenomena, which survives only in Latin. As he wrote:

> I am well aware that we live in an age when men of judgment consider it their part to greet a report of supernatural phenomena with contempt and derision, particularly when these phenomena are spiritual apparitions and the unusual communication of rational creatures with rational creatures that do not belong to the human race.

He went on to spell out how 'probable accounts of the phenomena of the invisible world' supplied by travellers and others might be useful partly 'if the recounted phenomena are extraordinary rather than supernatural', in which case 'the naturalist will more thoroughly investigate their cause, and a new philosophical light will dawn', but partly because:

> if he observes some phenomena that are above nature, there will arise the humbler consideration that there are objects beyond the grasp even of a philosopher in this life, and that some truths are not explicable by the powers of matter and motion, which truth is indeed of great importance in this age, when the Epicureans use the notions of their philosophy to reject everything that is contrary to it.[57]

On the other hand, he was only too aware that the scepticism he deprecated had been encouraged by cases of fraud and by undue credulity on the part of those who had recorded such stories in the past. 'Nor am I surprised that men

---

[56] MacIntosh, *Boyle on Atheism*, p. 338.

[57] *Scrupulosity and Science*, pp. 230–31. In Royal Society MS 191, fol. 110, appears the following passage, perhaps a trial version for this: 'I am sensible that I write in an age wherein som men of parts & many that hope by an affected severity to be reputed such, are very forward to distrust, if not condemn all Relations wherin any thing is deliver'd that supposes or may argue the being or Intervention of some Intelligent Agent that is not human' ('be reputed' is inserted, replacing 'pass for' deleted, and after 'supposes', 'the be' has been written above the line).

of penetrating judgment are not so easily led to give credence to wonders,' he wrote, 'given the various deceptions of impostors and the ready assent these deceptions gain from simpler men, as well as the eagerness of humans to impose their own fabrications on others.' Hence he consciously distanced himself from Aristotle in this respect, claiming that his collection of 'Strange Reports', unlike that of his ancient predecessor, would eschew 'groundles traditions or popular rumors' in favour of particulars derived from 'men of credite & reputation'.[58] We have already seen how Boyle often invoked fraud as a likely explanation of pagan 'miracles', and he made a similar point in his exchange of letters about witchcraft with Glanvill in 1677–78, when he urged Glanvill to be:

> very carefull to deliver none but well attested narratives. The want of which Cautiousnes's has justly discredited many Relations of Witches & Sorceries & made most the rest suspected since in such stories, the number of the whole can no way compensate the want of truth or of proofe in some of the Particulars, & a few narratives cogently verifyd will procure greater credit to the cause they are brought to Countenance than a far greater number of Stories where of some tho never so few are false, & others tho perhaps not many suspitious.[59]

This being the case, it is not surprising to find Boyle being particularly insistent on the need for verisimilitude in the accounts he recorded. This concern about testimony on his part has, of course, been discussed by Steven Shapin, but it is worth stressing that phenomena like these raised particularly acute issues of reliability which could not be resolved simply by recourse to considerations of status.[60] In his interview with Pierre-Esprit Radisson, for instance, Boyle 'let him know the things I desyrd were not uncertain reports, but candid accounts of what he himself had observ'd or receivd from persons whom he himself beleevd'. Similarly, with the Swiss captain whom he quizzed on similar topics, he insisted on 'personall knowledge', while he found a report concerning Indian snake charmers the more convincing because it was from an informant 'not forward to admit Witchcrafts or Inchantments'.[61] An interesting ancillary instance concerns a report concerning an alchemical adept in Italy who claimed to be 173 years old, on which Boyle comments:

> which relation thô scarce credible I was the lesse disposd to reject because the person I had it from seemd to be noe Charleton but a plain honest German of good repute amongst some of my acquaintance that have knowne him here & a

---

58 *Works*, vol. 11, p. lxviii.
59 *Correspondence*, vol. 5, p. 20.
60 See Steven Shapin, *A Social History of Truth* (Chicago, IL, 1994), esp. chs 5–6.
61 Workdiary 36–36, 102, 103.

man that in divers discourses I had with him seemd carefull not to affirm things that he had not tryed or did not otherwise know to be true; nor did hee at all pretend to bee acquainted with any of this Artists secrets for the prolongation of life.[62]

Equally interesting is the strongly critical tone that Boyle adopted in the course of these interviews. The workdiaries are particularly revealing from this point of view, giving a sense of the dialogue that ensued between Boyle and his informant, in which Boyle's interjections might at times almost make him sound like 'a Sceptic' himself (here, I echo his usage in another workdiary, where he cites objections that 'might by a Sceptic be pretended' as a reason for repeating an experiment[63]). What is striking is how alert Boyle was to the possibility of fraud. It is interesting that, in his preamble to *Experimenta & Observationes Physicæ*, he emphasised how he had recorded his experiments 'truly and candidly, without fraudulently concealing any part of them, for fear they should make against him'.[64] In his narrative of the Swiss poltergeist, he questioned whether a great stone that fell into 'a faire Roome' at the castle where it occurred without making any impression on the floor was 'a fantasticall whime', which disappeared again 'after the turn was servd'. Similarly, in the case recorded by Radisson, he spoke of how the American Indian priest 'playd his tricks' in invoking the spirits.[65] With Richard Cony, he opened his account by recording Cony's provision of a naturalistic explanation for a strange phenomenon that had perplexed the natives, 'from which it is clear that our colonel is not an excessively credulous type', prior to turning to 'the affirmative part of his answer'.[66] Yet, as this last instance reveals, for Boyle a show of scepticism was a means to an end rather than an end in itself, in that the bottom line was an attempt at verisimilitude in reporting such cases of supernatural interference in the world as a means of convincing those who were inclined to reject them altogether.

What, therefore, did the stories Boyle recorded entail? The phenomena recorded were bizarre. The Swiss case concerned a kind of poltergeist at a castle called La Chaux about eight leagues from Geneva, to the owner of which Boyle's informant was related. This involved 'severall strang noyses that could not <as was conceivd> be accounted for upon naturall grounds' and such 'odde things' as the great rock falling into a room without damaging it to which I have already referred. When asked by Boyle if he had seen any 'extraordinar sight', 'he replyed

---

[62] Workdiary 36–90.
[63] Workdiary 38–4.
[64] *Works*, vol. 11, p. 371.
[65] Workdiary 36–102, 103.
[66] *Scrupulosity and Science*, pp. 245, 249.

that he had not seen anie spirit appear in a visible shape, but that he had seen some performances that must have been made by some spirit'. For instance:

> when a house maid that lookt to the dishes & plates went about to wash them according to the custome of the countrey when they were carried down from off the table, the spirit would sometyms in a trice range them in order in the places they were to be put in of which surprising prank he himself was an eyewitnes.[67]

As for the ethnographic data supplied by Cony and Radisson, as also comparable information that Boyle garnered from the French savant François Bernier, these focused on native beliefs in Africa and Canada concerning the invocation of supernatural beings. Cony described a strange cloud formation which seemed to be induced by demonic activity. With Bernier and Radisson, in both cases voices of demons were heard making predictions which subsequently turned out to be accurate. Radisson described the erection of a kind of structure into which people went to invoke such beings. On one occasion, Radisson heard 'a very loud noise ... which seemd to be a confusion of the noises of I do not know how many severall sortts of beasts'; he also noted a man whose body was only partly inside the structure, so that his buttocks and legs protruded, which were the subject of 'violent commotions & contortions':

> When the savage came out of this dark place my relator observd his countenances [*sic*] to be quite alterd & disfigurd especiallie by a change of featurs that made him look very Ghastly <his> under lip hanging down allmost upon his chinne & his under eyelids being quite turnd up[:] he was allso for a pretty while unable to speak to him, but by the help of time & a pipe of Tobacco he recoverd his speech & told them severall things that I do not now remember relateing to their affairs[;] but in short my Relator as little credulous as he is affirmd that upon the whole <matter> he was convincd that there was something supernaturall in what he saw & heard.[68]

What is one to make of this? Of course, one cannot be certain which of these stories Boyle might have considered sufficiently convincing to publish, though the fact that the accounts derived from Cony, Radisson and Bernier were translated into Latin along with the preface to the unpublished Part II of 'Strange Reports' suggests that Boyle *was* satisfied enough with them to put them into circulation. It is also unclear how Boyle could be sure that these phenomena were genuinely supernatural, particularly in the light of his scepticism about the invocation of witchcraft as an explanation, or about the analogues to miracles

---

[67] Workdiary 36–103.
[68] *Scrupulosity and Science*, pp. 245–6, 247–8, 249–50; Workdiary 36–102.

in non-Christian religions, to which I have already referred. In any case, as I have indicated elsewhere, the project ran into complications, in that Boyle ultimately withheld the 'supernatural' part of 'Strange Reports' for reasons of 'discretion' – evidently in deference to the growing distaste for claims of this kind in late seventeenth-century English culture that I have already surveyed.[69] The fact that he had to wait till 'our Company withdrew' to ask Raddison about such phenomena is a further instance of his susceptibility to such pressures.

Be that as it may, what is surely significant is Boyle's active concern to accumulate empirical data proving the reality of supernatural phenomena, as also seen in his involvement in Glanvill's witchcraft project – where Glanvill seems to have supplied data to Boyle at his request as well as vice versa – or in such other initiatives on Boyle's part as his collection of instances of 'second sight' in Scotland in the late 1670s.[70] Indeed, the evidence suggests that his sense of the importance of this activity may have increased in his last years, particularly from the late 1680s onwards.[71] There is thus no equivalent to these insistent enquiries about the 'supernatural' in earlier workdiaries which are otherwise comparable to those in which this data appears, and we do not even know whether the original version of 'Strange Reports' that appears in a list of Boyle's writings around 1670 was to include 'supernatural' as well as 'natural' phenomena, as the later one did.[72]

This change was almost certainly linked to an increasing sense of the need for apologetic on Boyle's part. His awareness of the growing threat of religious heterodoxy in his later years is indicated, for instance, by the poem asserting the virtues of revealed as against natural religion that he wrote in response to a 'Deist', which was published by his protégé David Abercromby in 1685.[73] It was also in these years that – probably at the behest of his confidant Bishop Gilbert Burnet – Boyle formulated the idea of a lecture series 'for proveing the Christian Religion against notorious Infidels (*viz*ᵗ) Atheists, Theists, Pagans, Jews and Mahometans' which was to be implemented under the provisions of his will in

---

[69] *Scrupulosity and Science*, ch. 10.

[70] See esp. *Correspondence*, vol. 4, pp. 455–6, where the initiative seems to be Boyle's, in contrast to that taken by Glanvill from pp. 460–61 onwards. See also Michael Hunter (ed.), *The Occult Laboratory: Magic, Science and Second Sight in late Seventeenth-century Scotland* (Woodbridge, 2001), pp. 6–10 and 51–3.

[71] It is revealing that the lists of 'supernaturall phænomena' quoted in *Scrupulosity and Science*, pp. 228–9, are from notebooks dating respectively from *c.* 1690 (RS MS 186) and 1690–91 (MS 187). See also the dates of the interviews quoted on pp. 163–4 above.

[72] See above, pp. 176–7. For comparable earlier workdiaries see esp. Workdiaries 21 and 24, dating from the late 1660s/early 1670s.

[73] *Works*, vol. 10, pp. lviii–lxi.

the form of the famous Boyle Lectures that were to play such a central role in orthodox polemic in the following decades.[74]

Hence it seems to me that various conclusions may be drawn. The first is Boyle's evident ambivalence about the supernatural, his proclivity to maximise the remit of the natural in certain contexts while insisting on the reality and significance of the supernatural realm in others. As I have indicated, this has been obscured by certain historians who have taken Boyle to be in the vanguard of the increasing tendency to treat all phenomena as natural, neglecting his commitment to the supernatural. On the other hand, in this they were reflecting the impression given by Boyle's published works, since his most significant writings on such topics – on miracles as well as on strange phenomena – remained unpublished until modern times. This suggests a further conclusion, namely that it is important for us to understand the discrepancy that might exist between Boyle's aims and their implementation, as seen in this case by his initial withholding from publication of a compilation which he clearly considered important, evidently in response to trends in the culture of his period of which he was more aware than his opaque references in his published writings might make it appear.[75] It is a further paradox that the same cultural forces that discouraged him from publishing the book had encouraged him to compile it in the first place – in that Boyle's concern to collect empirical accounts of overtly 'supernatural' phenomena seems to have intensified in his later years in response to these very naturalising trends. Hence the picture that emerges is a complicated one, with various forces in operation which left Boyle in a typical state of convolution. As so often, by doing justice to this, we arguably come to the heart of Boyle's intellectual personality.

---

[74] See *Boyle by Himself and His Friends*, pp. xxiv–xxv, lxxxiv–lxxxv; R.E.W. Maddison, *The Life of the Hon. Robert Boyle* (London, 1969), pp. 274–5. On the significance of the Boyle Lectures, see M.C. Jacob, *The Newtonians and the English Revolution 1689–1720* (Hassocks, 1976); Michael Hunter, *Science and Society in Restoration England* (Cambridge, 1981), ch. 7.

[75] See the discussion in *Scrupulosity and Science*, ch. 10. However, since Boyle's programme for *Experimenta & Observationes Physicæ* was truncated by his death, it is quite possible that, as with 'Physica Peregrinans', the second part of 'Strange Reports' might have formed a component of a later volume had he lived, since it is indeed referred to in various of the lists of putative contents referred to in *Boyle Papers*, p. 187: cf. below, p. 185.

Chapter 9

'Physica Peregrinans, or the Travelling Naturalist': Boyle, His Informants and the Role of the Exotic

The title of this chapter alludes to a work planned by Boyle in his later years. 'A Catalogue of the Honourable Mr. Boyles Writings unpublish'd taken the 7th of July 1684' includes an item described as: 'A pretty large collection of things that I learn'd in conversing with Pilates, Sea-captains, and other Persons that have travail'd unto the Indies, or other remote Country in Order to make Physica peregrinare'.[1] It transpires that this was to be a component of the miscellaneous compilation that Boyle began to publish right at the end of his life, his rather inelegantly titled *Experimenta & Observationes Physicæ, Wherein are briefly Treated of Several Subjects Relating to Natural Philosophy in an Experimental Way*.[2] The only volume of this to be published appeared in July 1691, five months before Boyle's death; it included various components, including groups of magnetic and chemical experiments, notes on diamonds and colour changes, and medical cures. At the end was the first part of a separate section entitled 'Strange Reports', which gave narratives of various phenomena which were puzzling but nevertheless clearly natural; as we saw in Chapter 8, a sequel to this was planned dealing with 'supernatural' phenomena which Boyle initially held back, though there are some clues to its intended content. From various documents among Boyle's papers with titles like 'A Scheme or Method of the Experimenta & Observationes Physicæ', it is clear that the published volume was meant to be only the first of a series, comprising 'severall Parts, or distinct Tomes because they are not intended to be publisht all at once'.[3] Among the components itemised in various of these lists, destined for the second or third volume, is the work: 'Physica Peregrinans, or the travelling Naturalist[,] containing Answers given to Severall Questions propounded by the Author to Navigators & other Travellers into remote Countreys'.[4]

---

[1] *Works*, vol. 14, p. 342. Cf. ibid., p. 347, a document included in a list of papers dated 22 July 1688 described as 'Relations of Travellers broad 4° parchment'.

[2] See ibid., vol. 11, pp. liii–liv, lxi–lxii, 367ff. See also above, pp. 176ff.

[3] See the discussion in *Boyle Papers*, pp. 186–8.

[4] BP 36, fol. 84. For other versions, see RS MS 186, fols 28v, 122v: the former pre-dates both the latter and the version in BP 36, fol. 84, being emended in composition with

This compilation, the material that would have gone into it, and Boyle's rationale in composing it, will form the subject of this chapter. It is clear that the principal source for the otherwise lost 'Physica Peregrinans' would have been the lengthy series of accounts of interviews with seamen, travellers and government and colonial officials that form part of Boyle's workdiaries.[5] As we saw in Chapter 2, after initially comprising extracts from the French romances with which Boyle was so enamoured as an adolescent, in 1649 the workdiaries provide a dramatic testimony to his conversion to science, since at that point literary extracts are replaced by records of recipes, experiments and other natural philosophical data, and this continues until Boyle's death. However, there is some variety in the content of the later workdiaries, and among the most appealing are two, Workdiary 21, a collection of 'Promiscuous Experiments, Observations & Notes' dating from the late 1660s and the early 1670s, and Workdiary 36, dating from between 1685 and Boyle's death: both are predominantly devoted to recording interviews Boyle conducted with a variety of informants, many of whom had travelled to exotic locations in different parts of the world. Indeed, the former has its original vellum cover, reproduced in this volume as Plate 8, which is tellingly entitled 'The Outlandish Booke'.[6] There can, I think, be no doubt that it was to the interviews recorded in these workdiaries, and probably also in other comparable ones that are now lost, that Boyle was referring in the subtitle to 'Physica Peregrinans'.[7] This was presumably intended as a kind of anthology of these interviews, thus evidently bearing witness to Boyle's sense of the value of such records in their own right, something to which I will return later in this chapter. But it is clear that, when he initially carried out the interviews and recorded them in his workdiaries, Boyle's primary aim was more directly utilitarian – of adducing reliable data about unusual phenomena, or phenomena in remote regions that he could not experience for himself, which would provide key evidence concerning the workings of the natural world and the manner in which these were to be explained.

'given' inserted in place of 'made' and 'Questions propounded' replacing 'Enquirys made'. In BP 36, fol. 19 (another synopsis of the content of *Experimenta & Observationes Physicæ*), the work is just called 'Answers & Relations made me by Navigators, & other Travellers'. For a hint that Boyle already had the idea of such a compilation much earlier, see the mnemonic list of his writings of *c.* 1670, no. 34 of which *'shew's the use of travells great'*: *Works*, vol. 14, p. 336. It is almost certainly not coincidental that this document survives in juxtaposition with Workdiary 21: see ibid., p. xlii.

[5]	http://www.livesandletters.ac.uk/wd/index.html. For an overview, see *Boyle Papers*, ch. 3.

[6]	Above this are the letters 'AA', perhaps relating to Boyle's system for filing his papers: see *Scrupulosity and Science*, pp. 121ff., and *Works*, vol. 14, pp. 329ff.

[7]	For lost workdiaries, see *Boyle Papers*, esp. pp. 167ff. See also Appendix 2 to this chapter.

The inspiration for this was the imperative Boyle had inherited from Francis Bacon for an accurate natural history as the basis for natural philosophy. Indeed, in the 'Advertisements' for *Experimenta & Observationes Physicæ*, the work of which 'Physica Peregrinans' was to form part, Boyle provided an elaborate Baconian rationale for utilising data supplied by others, claiming that he was:

> Authoris'd in that Practice, by frequent Examples afforded me by the first, if not only Author that I know of, that gave us a Set of Precepts of well writing Natural History, our often cited *Verulam*, whose Centuries do in great part consist of borrow'd *Experiments* and *Observations*; without which, he was sensible that his *Sylva* must be of too narrow a compass, or too thinly stock't with Plants, especially with Trees. And indeed 'tis not to be expected, that, as the Silk-worm draws her whole Mansion altogether out of her own Bowels, so a single man should be able to write a Natural History out of his own Experiments and Thoughts.[8]

In many ways, this echoed the 'Designe' for the proper method of natural history that Boyle had drawn up in the 1660s, in which he had envisaged the testimony of others as playing an important role in supplementing his own experimental findings, particularly in connection with such things 'as are not to be examin'd but in remote Countrys, or Places we cannot come to'.[9]

This issue had arisen with particular intensity in his *New Experiments and Observations Touching Cold*, published in 1665, since in that work Boyle wanted to comment on extreme weather conditions found in remote parts of the world that he had not visited himself, and his strategy in it has been the subject of particular attention from Steven Shapin in his *Social History of Truth*.[10] Primarily, Boyle's method was to read books, most of them by voyagers who had explored the polar regions a generation or more before his own, including the Dutch sailor Gerrit de Veer, whose 1598 account of his voyages was reprinted in *Purchas his Pilgrimes* in 1625, and especially Thomas James, whose *Strange and Dangerous Voyage ... in his Intended Discovery of the Northwest Passage* (1633) Boyle cited more than any other work. In the preface to *Cold*, Boyle provided an elaborate justification for his reliance on such authorities, going further than in any other of his writings in providing a rationale for his use of books in this

---

[8]   *Works*, vol. 11, p. 375.

[9]   Peter Anstey and Michael Hunter, 'Robert Boyle's "Designe about Natural History"', *Early Science and Medicine*, 13 (2008), 83–126, on pp. 107–9; Michael Hunter and Peter Anstey (eds), *The Text of Robert Boyle's 'Designe about Natural History'*, Robert Boyle Project Occasional Papers No. 3 (London, 2008), p. 4.

[10]   Steven Shapin, *A Social History of Truth: Civility and Science in Seventeenth-century England* (Chicago, IL, 1994), pp. 247ff. Boyle also made some use of travel books in his *Experiments and Considerations touching Colours* (1664).

connection, not least in terms of seeking to assess the credibility of their authors.[11] Boyle also exemplified more fully in *Cold* than in any other work a preferred method of citation which he justified in his 'Designe about Natural History', written shortly afterwards, namely of noting exactly which edition of the book in question had been consulted and recording the exact page on which a cited passage appeared – a practice which has made it possible partially to reconstruct his otherwise lost library.[12]

In *Cold*, therefore, Boyle was predominantly dependent on books for his information, and he continued to use books, and not least travel books, thereafter – especially *Purchas his Pilgrimes*, an extensive body of extracts from which survives in Volume 39 of the Boyle Papers, along with notes from travel books by J.B. Du Tertre, Richard Jobson and others.[13] Boyle appears to have valued his notes on Purchas highly, even having his amanuenses emend them with elucidations and connecting passages in order to make them into more of a reading text and to facilitate reference to the original works Purchas anthologised, as Iordan Avramov and I have shown in a broader study of Boyle as a reader.[14]

Boyle occasionally quoted oral sources as well as printed ones in *Cold*, but his use of this technique was there limited.[15] In subsequent years, however, this practice was to come to the fore, as seen not least in the workdiaries and the writings in which he drew on them, and the background to this needs to be sketched here. One aspect of this is Boyle's discovery of the value of questionnaires as a means of eliciting data from living informants, and not least from those who travelled to exotic places. As we saw in Chapter 3, this was a practice which Boyle developed in conjunction with the Royal Society in its early years, when the society seems initially to have taken the initiative more than Boyle himself.[16] But Boyle soon made the genre his own, producing perhaps the classic example of it in his 'General Heads for a *Natural History of a Country*, Great or small', published in *Philosophical Transactions* in 1666; indeed, the significance of this

---

[11] *Works*, vol. 4, pp. 217–21. For a commentary, see Anstey and Hunter, 'Boyle's "Designe"', pp. 108–9, and Iordan Avramov and Michael Hunter, 'Reading by Proxy: The Case of Robert Boyle (1627–91)', *Intellectual History Review*, 25 (2015), 37–57.

[12] See Hunter and Anstey, *Text of Boyle's 'Design'*, p. 2; Iordan Avramov, Michael Hunter and Hideyuki Yoshimoto, *Boyle's Books: The Evidence of His Citations*, Robert Boyle Project Occasional Papers No. 4 (London, 2010), pp. xvii–xviii and passim.

[13] BP 39, fols 1–48, 53–73, 81–94.

[14] BP 39, fols 1–48; Avramov and Hunter, 'Reading by Proxy'.

[15] See, for example, *Works*, vol. 4, pp. 219, 220.

[16] See Chapter 3 in this volume. For a recent full account of the background, see Daniel Carey, 'Inquiries, Heads, and Directions: Orienting Early Modern Travel', in Judy A. Hayden (ed.), *Travel Narratives, the New Science and Literary Discourse, 1569–1750* (Farnham, 2012), pp. 25–51.

document is underlined by the fact that Boyle gave numerical codes that can be linked to it to many entries in Workdiary 21, the first of the compilations in which travellers' data is prominent.[17] Boyle also compiled a whole series of questionnaires for his own private use, many of which still survive.[18] It is therefore hardly surprising to find that in the records he kept in his workdiaries of the interviews he conducted, as in the title of 'Physica Peregrinans', Boyle repeatedly refers to the 'questions' he asked of his interlocutor and the 'answers' he received, in at least some cases recording the latter in a numbered sequence, as we will see.

In fact, at least two sets of inquiries of this kind survive. One is a set of queries aimed at a specialist in the refining of lead, presumably somewhere in England, which comprises various detailed questions about the quantities and methods used in the process.[19] The other, entitled 'Quæries for Mr Dabers', concerns sailing times and seasons and wind patterns, and it gives a good sense of the informal questionnaires Boyle must have compiled, the content of which can otherwise only be deduced from the replies he received. The 'Quæries' read as follows:

> In how long time the Vessels that goe hence for Suratte doe usually double the Cape of Good Hope & in what time they reach from thence to Suratte, Bombaijm [Bombay] & other places on the Coast of Cormandell, as also what seasons & spaces of time Voyages are wont to be made to Bantam, Jamby, Maccassar & other Southern places we trade too. What trade winds & other winds doe they meet with in their way & in what places & seasons of the year.[20]

The advantage of questionnaires over reading was that they could be pro-active: in other words, instead of having to try to find out what you wanted from the narrative of an author whose preoccupations might have been quite different – as had been the case with Boyle's use of accounts of voyages in *Cold* – it was possible to explain exactly what information was required and to solicit this from someone who might otherwise not have thought it worthy of record. As we will see, this is often precisely what happened in Boyle's interviews. Moreover, as with the questionnaires which Boyle and his Royal Society

---

[17]   See *Works*, vol. 5, pp. 508–11, and above, p. 65.

[18]   See above, pp. 61ff., and Michael Hunter (ed.), *Robert Boyle's 'Heads' and 'Inquiries'*, Robert Boyle Project Occasional Papers No. 1 (London, 2005), passim.

[19]   BP 10, fol. 49.

[20]   BP 39, fol. 199 ('Bomaijm' is followed by 'as al' deleted; in the following line, 'also' is followed by 'att' deleted and 'what' by 'tim' deleted). 'Mr Dabers' is conceivably George Davies, who was Purser on the *Vine*, which sailed to Surat under Captain Edward Mason in 1658–59: see Anthony Farrington, *A Biographical Index of East India Company Maritime Service Officers 1600–1834* (London, 1999), p. 205, and Anthony Farrington, *Catalogue of East India Company Ships' Journals and Logs 1600–1834* (London, 1999), p. 679.

colleagues had compiled earlier, the questions he formulated were often based on extensive reading in books on related subjects, which set him an agenda of puzzling phenomena about which further information was required.[21] There are thus cross-references to travel writers like Samuel Purchas, Adam Olearius or Gonzalo Hernández de Oviedo y Valdés, to mineralogical writers like Jean-Baptiste Morin or to natural historians like Conrad Gesner.[22]

Hence, by comparison with the method exemplified by *Cold*, this face-to-face technique of interrogation had distinct advantages, as Boyle pointed out more than once after he had discovered the new approach. Thus the appendix that was added to the second edition of *Cold* published in 1683, which largely comprised material of this kind, was prefaced by a note by Boyle's 'publisher', J.W., which combined a Baconian justification for the use of books with an equal emphasis on using interviews:

> It remains that the Reader be told, whence the Materials have been taken, whereof the following *Appendix* doth consist. Some few of them have been drawn out of Printed Books, because *Cold* (in it self a subject barren enough) has been left so uncultivated by Classick Authors, that, according to our judicious *Verulam's* advice, it was not thought fit to cast away any credibly related matter of fact that might add to the History of it. But the greater number by far of the following Particulars was taken from the Relations of Navigators and Travellers, whom the Author had the Curiosity to consult about the *Phænomena* of *Cold*, they had met with. And for the better gaining of such Informations, he became an Adventurer in that which is commonly called the *Company of Hudson's Bay*; to which those that are from time to time sent from *London*, do, either in their Voyages thither and back again, or in their stay in that frozen Country, not unfrequently meet with considerable, thô unwelcome effects of *Cold*.[23]

A similar statement appears in Boyle's 'Observations and Experiments about the Saltness of the Sea', part of the volume of *Tracts* he published in 1673, which contains the following 'Advertisement To the following Observations (which may also serve for many Historical passages in the Author's other Writings)'. In it, apart from a caveat about reliability, Boyle comparably capitalised on his unprecedented use of oral information:

---

[21]   For the Royal Society, see above, pp. 70, 72; Michael Hunter, *Establishing the New Science: The Experience of the Early Royal Society* (Woodbridge, 1989), pp. 93–4, 118–20, and Carey, 'Inquiries, Heads, and Directions', pp. 46–7.

[22]   Workdiary 21–372 (Oviedo), 421 (Morin), 550 (Olearius), 675a (Purchas); Workdiary 36–2 (Gesner). See also Workdiary 21–308 for a reference to Bacon.

[23]   *Works*, vol. 4, p. 546. 'J.W.' was Boyle's assistant, John Warr.

Whereas the Author does frequently make use of the Relations of professed Seamen and other Navigators, and of Observations made some in the *East*, and some in the *West-Indies*, it will be fit to advertize the Readers, that he has been very wary in admitting the informations that he imployes; being forward enough to reject, as he has often done, such as many others would gladly have received: But notwithstanding his wonted rejection of the particulars he saw cause to disbelieve, 'twas easie for him to be well furnished with such relations as he makes use of; scarce any Writers of Philosophical things having had such opportunities of receiving such Authentick Informations from Sea Captains, Pilots, Planters, and other Travellers to remote parts, as were afforded him by the advantage he had to be many years a member of the Council appointed by the King of *Great Britain* to manage the business of all the *English Colonies* in the Isles and Continent of *America*, and of being for two or three years one of that Court of Committees (as they call it) that has the superintending of all the affairs of the justly famous *East-Indian Company* of *England*.[24]

These passages provide another clue to the background to these interviews, namely Boyle's move of his main place of residence from Oxford to London in 1668, and the activities on his part this facilitated. Obviously, he had spent a good deal of time in the metropolis prior to this, and it seems likely that he was occasionally able to talk to travellers and seamen at an earlier date: in fact, as he himself notes in the passage quoted above, Boyle had been a member of the Council for Foreign Plantations in the early 1660s, and it may initially have been this that made him aware of the opportunities the contacts associated with such bodies presented.[25] But it seems likely that, once ensconced on a virtually permanent basis in Pall Mall in 1668, it was easier for Boyle to summon travellers and others for interview than had previously been the case. It is also significant that it was at this time he became a 'committee' of the East India Company, as he noted in 'Saltness of the Sea', and hence was able to collect information from its officers, and the same was true of his enrolment in the Hudson's Bay Company in 1675, to which J.W. refers in the 1683 *Cold* appendix.[26]

---

[24] Ibid., vol. 7, p. 390. For a further reference to the use of the evidence of travellers see the preface to *General History of Air*: ibid., vol. 12, p. 10.

[25] See *Between God and Science*, p. 129. Note also Boyle's comment quoted on p. 74 above. For a rare example of a comparable interview in an earlier workdiary, see Workdiary 19–22 (Cholmeley on Tangier); other reportage is all from informants in England (for example, Workdiary 15–52, 19–7). A statement by Boyle at a meeting of the Royal Society on 23 September 1663 seems to reflect such an encounter: see Birch, *Royal Society*, vol. 1, p. 305 (cf. ibid., p. 329).

[26] *Between God and Science*, pp. 169–71, and ch. 10, passim. It is possibly significant that a lengthy series of responses by voyagers to Hudson's Bay to inquiries furnished by

Also relevant is the kind of writing Boyle was doing at this time. As we saw in Chapter 7, in the early 1670s Boyle's characteristic writing style changed markedly, away from the substantial treatises he had published in the early to mid-1660s, of which *Cold* was a classic example, to briefer, more wide-ranging and more speculative treatises.[27] Typical examples of these comprise his *Essays of Effluviums* (1673) and the volume of tracts including 'Saltness of the Sea' that has already been cited, along with essays on the nature and virtues of gems, on the temperature of the inner regions of the earth, on the nature of the seabed or on the 'hidden qualities' of the air, not least in affecting minerals. Obviously, Boyle also remained interested in more general questions of the kind he had addressed in the books he had published in the 1660s, for instance concerning the nature of heat and cold, or the role of air in respiration or in creating abnormal atmospheric conditions. He also had a programmatic interest in the vindication of the corpuscular philosophy, while his longstanding concern with the potential utility of science made him curious about the technology of other cultures, such as China.[28] He was also interested in the flora and fauna of different geographical areas and the variations between these and those familiar in Europe, partly in their own right, and partly in the hope that such exotics could be profitably introduced at home. Indeed, as we saw in Chapter 3, the latter was the theme of an essay for Boyle's *Usefulness of Natural Philosophy* which is now sadly lost.[29] Hence these are the kinds of components one would expect in Boyle's agenda at the various meetings he had. On the other hand, it is perhaps worth noting here the sort of topics on which travellers would surely have been happy to report yet in which Boyle seems to have shown virtually no interest, particularly human customs and institutions, and antiquities.

The Interviews

Turning now to the interviews themselves, and to Boyle's record of them, various points may be made. They seem mostly to have occurred in Boyle's rooms in his sister Lady Ranelagh's house in Pall Mall: on one occasion, Boyle specifically states that the meeting took place in 'the chamber <next to that> I ly in at London', and they probably normally occurred in his 'great Room', the larger of

---

Oldenburg had been read at a Royal Society meeting on 18 April 1672, which might have given Boyle the idea: Birch, *Royal Society*, vol. 3, pp. 43–6.

[27] See above, pp. 154–5.

[28] For technology, see, for example, Workdiary 21–419 (varnish), 565 (mending china). For the other concerns, see Workdiary 21, passim.

[29] See above, p. 74. Cf. his interest especially in the ways in which New England 'differs from other Countrys, especially those we here live in' in his enquiries about its natural history: see *Correspondence*, vol. 2, p. 462 (and see ibid., pp. 268, 457).

the suite of two rooms he occupied.[30] Most often, the meetings appear to have been with a single individual, but occasionally the presence of others is noted.[31] Notes on the conversations were then dictated to one of the amanuenses Boyle habitually employed, and this usually seems to have occurred fairly soon after the event. We learn from a later memoir of Boyle by Thomas Dent, chaplain to his brother, Lord Shannon, how Boyle daily spent time 'entertaining persons of severall nations' and how 'what was remarkable in Experiment or occurrence, he note[d] down <Every day> when the company departed'.[32] The fact that notes on two interviews are sometimes intertwined could be due to the fact that they took place on the same day, though it is equally possible that recording was not always as immediate as Dent's comments imply and that notes on a few days' worth of interviews might sometimes have been entered together: that this was the case is suggested by places where Boyle could not remember certain details of what he had been told.[33] In Workdiary 21, the interviews appear among records of experiments, presumably because it represents a sequential account of different activities on Boyle's part, and in one case a single interview is interspersed by an account of an experiment: this was possibly because it occurred on the same day, or it might even have been demonstrated to the visitor whose comments are recorded before and after it.[34] On the other hand, Boyle seems to have kept Workdiary 21 in parallel with Workdiary 22, which is principally devoted to extracts from books, and with other workdiaries of comparable date, and a similar practice is evident in connection with Workdiary 36, which is devoted exclusively to interviews and which was apparently kept in parallel with Workdiaries 37 and 38, in which accounts of experiments from the same period appear.[35]

---

[30] Workdiary 21–706; *Between God and Science*, p. 166.

[31] For example, Workdiary 21–748; Workdiary 36–34, 76. See also below, p. 201.

[32] *Boyle by Himself and His Friends*, p. 105. Dent emphasises the mornings as the time for such visits, perhaps because that was when he himself visited Boyle, whereas Evelyn in ibid., p. 89, stresses the afternoon as the time when visits predominantly took place.

[33] For intertwined interviews, see Workdiary 21–340ff., 368ff., 537ff., 646ff. For details that Boyle could not remember exactly, see, for example, Workdiary 21–343, Workdiary 36–98, 100, 102.

[34] See Workdiary 21–270 and 272: the intervening entry, Workdiary 21–271, concerns acids, alkalis and colour changes.

[35] This is apparent from the dates included in the headnotes to entries in the workdiaries in question. See also *Boyle Papers*, pp. 159–60, 162. Other workdiaries that were probably kept in parallel with Workdiary 21 include Workdiary 23. The fact that a handful of entries relating to interviews with Henry Stubbe (see below, n. 37) appear in Workdiary 22–30ff. may be due to an error on the part of Boyle's amanuensis.

The interviews come to the fore at a specific point in Workdiary 21, namely in February 1668.[36] Perhaps ironically, the first interlocutor was the controversial figure Henry Stubbe, who was to cause problems for the Royal Society through his attacks on it over the next few years, though Stubbe throughout professed his respect for Boyle, who evidently relished the information Stubbe could give him about Jamaica, whence he had returned from a three-year stay in 1665.[37] These extensive interviews range in subject matter from atmospheric conditions to the flora and fauna of the island, illustrating Stubbe's interest in natural phenomena in the Caribbean and elsewhere, notwithstanding the reservations he felt about the Royal Society and its aspirations; indeed, such facets of Stubbe's activities deserve fuller study than they have yet received. The first of these interviews may have taken place in Oxford, prior to Boyle's permanent move to London in November 1668, but the intensity of interviews thereafter increases, doubtless reflecting Boyle's London location. Indeed, it is almost as if Boyle had a sudden 'craze' for collecting such data, since it seems likely that a high percentage of the extant records of this kind can be dated to 1668 or 1669, the first year when he lived continuously in London, thus also dovetailing with his composition of works on related topics, which started to be published within the next couple of years.

As for the identity of Boyle's informants, it is unfortunate that, both at this time and later, he is sometimes a little circumspect in identifying them, using opaque formulae which were evidently intended to be transcribed directly into his published books and which it is often impossible to elucidate – as with 'an observing person, not long since returnd from the Indies', or 'an antient Divine that spent divers years in America as a planter', or 'a great Traveller in both Indies'.[38] Indeed, it is almost as if such formulae were intended as knowing winks on Boyle's part to some circle of cognoscenti, and they are a far cry from the stress on named witnesses that has sometimes been seen as central to his intellectual strategy.[39] In certain cases, enough clues are available to make an informed guess at the identity of the individual involved, as with 'an Ingenious man that saild into the frigide Zone as farr as Greenland', who proves to be the

[36]   The chronology is slightly confused by the recopying of much earlier entries into the workdiary at this point, but it seems clear that the February/March dates in entries 248–9 and 262 relate to 1668.

[37]   See J.R. Jacob, *Henry Stubbe, Radical Protestantism and the Early Enlightenment* (Cambridge, 1983), p. 49 and passim. Stubbe is one of the most frequent interlocutors in the workdiaries, appearing in Workdiary 21–250ff., 284ff., 574ff., 616ff., and 719–20 (copies of entries 616 and 618). See also Workdiary 22–30ff., Workdiary 23–20, and *Boyle Papers*, pp. 170–71.

[38]   Workdiary 21–370, 414, 613.

[39]   See esp. Steven Shapin and Simon Schaffer, *Leviathan and the Air-pump: Hobbes, Boyle and the Experimental Life* (Princeton, NJ, 1985), pp. 57–8.

Hamburg surgeon Fredrick Martens, who made a voyage to Spitsbergen and Greenland in 1671, or 'an inquisitive Gentleman lately returnd from Jamaica' in May 1689, who turns out to be the young Hans Sloane.[40] On other occasions, Boyle actually gives his informant's name as well as some information about him, as with Sir Thomas Lynch, Governor of Jamaica, or the Earl of Sandwich (Plate 7), courtier and naval commander, who took the English fleet to Portugal to bring Catharine of Braganza to England to marry Charles II in 1661–62 and who was subsequently active at Algiers and elsewhere in the Mediterranean, to whose interviews with Boyle we will come shortly.[41]

There is also a group of entries relating to officers of the East India Company, including Captain Andrew Parrick, who sailed the frigate *Zant* to Madras and Bengal in 1669–70, and John Proud, who had commanded various ships in the 1630s and 1640s and was the company's surveyor of shipping in the 1660s and 1670s: indeed, these occur in a little clump, as if Boyle suddenly felt impelled to gather data on issues to do with the maritime conditions on which such men could inform him.[42] Overall, an extraordinary range of figures appear, including foreign informants as well as English ones – more than one from Portugal, and various Frenchmen, none of them very clearly identified, along with 'the Viceroy of Norway', in other words the statholder Ulrick Fredrick Gyldenløve, Danish Ambassador Extraordinary to England in 1669–70, who provided information about the efficacy of divining rods and about various aspects of mines in Scandinavia.[43] On the other hand, it has to be said that the selection is somewhat random, and it is unclear exactly how Boyle chose who to interview in this way: for instance, there were dozens of voyages by East India Company ships to the Far East at this time, and Boyle spoke to only a tiny percentage of those involved in them.[44]

If anything, those interviewed in Workdiary 36, from 1685 onwards, form a more predictable group, including many well-known travellers to exotic regions, whom Boyle often introduces with a flourish, obviously with the intention that the passage in question should be transposed verbatim into a published work. Thus the virtuoso traveller Sir John Chardin was introduced as 'a very candid & judicious Traveller', 'eminent for his Travels into Eastern Parts, & for his skill in Jewels', while Robert Knox, author of *An Historical Relation of the*

---

[40] Workdiary 36–87 (Sloane, cf. 104), 97 (Martens, cf. 91ff.).

[41] Workdiary 21–340ff. (Lynch), 312, 318ff., 520ff. (Sandwich). It is only in the last group that Sandwich is actually named, but it is a reasonable surmise that the earlier entries also relate to him. For the biographical details given here and hereafter see the 'Biographical register' at http://www.livesandletters.ac.uk/wd/index.html, based on Farrington, *Biographical Index*, and comparable works.

[42] Workdiary 21–633 and the entire series from 635 to 658.

[43] Workdiary 21–352ff., 548ff., 557ff., 579, 594, 681, 706ff. and passim.

[44] See Farrington, *Catalogue*, Appendix 1.

Island Ceylon (1681), is described as 'sufficiently known being the author of the History of Ceylon', and the French explorer Pierre-Esprit Radisson was said to be 'a Gentleman Captain whose skill in many things, & by severall voyages he has made to the Indies & in them, have procurd him considerable employments among two Emulous nations'.[45] At this time, Boyle also interviewed English colonial officers such as Richard Cony, Sir Thomas Rolt and Sir William Stapledon, former consular officials like Sir Paul Rycaut, and employees of the Hudson's Bay Company such as its Governor, John Nixon, and one of its captains, William Bond.[46]

The actual interviews that are recorded are striking for the vivid detail they include, as will be illustrated by extensive quotation in the course of this chapter, which it is hoped will encourage readers to investigate this rich body of material for themselves. As already noted, Boyle seems often to have prepared himself with a list of queries based on preliminary reading or investigation, and the data often comprise a set of answers to the questions he asked. Indeed, in the case of Robert Knox, Boyle seems actually to have given him a list of questions before he set out on the travels he undertook in the 1680s, interrogating him on them on his return.[47] The answers given are sometimes numbered, making it possible to reconstruct the questions that must have been asked, as in this instance from Workdiary 21, endorsed 'About spouts', and almost certainly aimed at the Earl of Sandwich, though, as often, Boyle's characterisation of his informant is frustratingly vague:

> Meeting with one of our English Admiralls <that is a virtuoso, & has> commanded many of the men of warr in the streights, & elswhere, I inquird of him whether he had mett with or taken notise of any of those strange meteors, that our seamen commonly call spouts, he replyd <To satisfie my Queries about particulars:> that besides those he had seene afar of[f] he once <saw one> so near at hand, that part of the water to a good quantity fell upon his ship. 2dly the Quantity of water that falls in one of these spouts is soe great, & may amount to so many Tun of water that if the Bulk of it should fall directly upon a ship: (As onely the edges of It did upon his) it would be capable of foundering & so, sincking the vessell, 3ly: That a litle before the fall he perceivd a Wind; perhaps made by Aire which the falling water drove before it 4ly that he saw the rising of one of this [*sic*] spouts out of the sea; which seemd to him to be afarr off, & to looke blackish, & not much bigger at first then a mans Thumb, but as it ascended upwards the higher

---

45   Workdiary 36–19, 78, 93 (cf. 6ff., 114), 102.

46   Workdiary 36–8ff., 23–4, 27ff., 53, 59ff., 68, 82–3, 92, 111ff.

47   Workdiary 36–93. For Robert Hooke's parallel dealings with Knox, see Lisa Jardine, *The Curious Life of Robert Hooke* (London, 2003), pp. 272ff., and Katherine Frank, *Crusoe: Daniel Defoe, Robert Knox and the Creation of a Myth* (London, 2011), esp. ch. 8.

[it] went the bigger it grew, till at length when he had followed it with his Eye till t'was out of sight it soone after reappeard in forme of a great darke cloud, upon which the seamen presently according to their wont made hast to take in their sayles, which they had scarce time to doe before the Cataract was pourd upon them 5ly That sometimes as farr as he could observe the rising of the spout, which not accompanyd with any or at lest with any vehement, or peculiar wind, as a whirle wind. 6th but after the falling of the spout, the weather was tempestious & some rainy for 2 or 3 howers 7th. that from the time he first descryed the rising steames to the falling of the water upon his ship he Estimated, that there had Effluxd between an hower or two or thereabouts. 8ly he added that he had not only seene this meteor in the Mediterranean, where 'tis observd to be not very frequent, but in the channel itselfe, but he found It more formidable then in the streights, & to be attended with furioser stormes.[48]

As will be seen, there are eight numbered answers, presumably reflecting the questions Boyle put, some seeking precise quantitative data, others reflecting a more general interest in the meteorological circumstances and ancillary weather conditions and the way these were juxtaposed.

A further instance in which numbered answers to Boyle's questions are given involves 'a Chymist that purposely visited the Mines in Hungary & Transylvania', probably (though not certainly) Dr Edward Browne, son of Sir Thomas Browne, and a noted physician and traveller in his own right, whose travels to central Europe were to be published in 1673.[49] On this occasion, Boyle's concern was with how the temperature varied underground, though, after itemizing three points, both Boyle and his interlocutor got slightly sidetracked by the issue of underground 'steames' which were said to be prognostics of storms above ground:

---

[48] Workdiary 21–312. See above, n. 41. A waterspout is in fact recorded by Sandwich in his journals under the date 4 August 1664: see R.C. Anderson (ed.), *The Journal of Edward Mountagu, First Earl of Sandwich, Admiral and General at Sea 1659–65* (London, 1929), pp. 146–7 (illustrated in the original MS of the journal, now at Mapperton House, Dorset, vol. 1, p. 218, by a sketch with the wind emanating from the mouth of an anthropomorphic figure that is omitted from the published edition). However, this must be a different one from that which Sandwich told Boyle about because (1) the 1664 instance was in the Channel rather than in the Mediterranean, and (2) none of the water from it fell on the ship: it is presumably one of the instances to which Sandwich refers under question 8.

[49] Workdiary 21–297ff. However, the chronology is slightly problematic and this may have been a different, otherwise unidentified informant: Browne was actually on his travels from August 1668 to Christmas 1669, whereas internal dating evidence suggests that this interview took place during or even before that time. For subsequent series which almost certainly *do* involve Browne, see Workdiary 21–368ff, 421ff. (in the latter, he is identified as 'Dr B.')

The forementiond Visiter of the Hungarian & other Mines, being askd by me about the Heat & Cold he found in the deepest of them answerd <to my severall Questions> First that as he descended he felt it cold till he came to such a depth as he had scarce attaind in a Quartar of an howers descent. 2dly That the cold he felt seemd to him considerable (except very near the Orifice of the Grove) especially when he got to a good depth. 3ly That after he had passd that cold Region he began by degrees to come into a warmer one, which increasd in heat, as he went deeper & deeper, so that in the deeper veines he found the workemen digging with only a slight Garment over them, & the Subterraneall heat was much greater then that of the free Aire on the Top of the Grove thô it were then summer, nor was it by any stifeling steames or want of free Respiration that they felt this heat. And as to those steames <(about which)> I inquir'd whether they observd them to be foreruners of stormes, He answerd me that the miners told him that even lesser damps were some Impediments to their Respiration, but when the thick & copious damps chancd to rise out of the Earth they were faine for fear of being stifeld to hasten out to the free Aire, & stay there till the Damp were gone, & if these damps were copious & thick enough, they would confidently & almost certainely foretell, that within a short time, there would be rainy or stormy weather even when the sky was clear & cloudles.[50]

In other cases, we get a more descriptive approach, though still with an insistent tone of inquiry on Boyle's part, as in this instance involving an unidentified informant who was able to provide data relevant to Boyle's strong interest in luminescent phenomena:[51]

A curious Physitian of my acquaintance that liv'd a while at Rochell related to me that heareing of some shineing Fishes that were wont to be seen about a place in the sea not very far off, he & some other Inquisitive Men caus'd themselves to be landed upon a certain sandy bank, which the sea did not quit cover which was near the above mentiond place. And this was purposely done in a dark but not stormy night; the success was, that he had the pleasure to see there a great number of Fishes playing up & down, that seem'd like so many Bodies moveing in the midest of the water, shineing not with the head or tail onely but the whole Body, as if they were so many liveing Carbuncles: To my question about the bignes of those fishes, he answerd, that they were about the size of those Sardinees which I had seen in the Ligustick Sea, & of which our Anchovies are made. <And> to my scruple whether the sea itselfe were not thereto dispos'd to shine so that the

---

50   Workdiary 21–299.
51   See *Between God and Science*, pp. 157, 167, 188–9. See also 'Mechanical Production of Light', *Works*, vol. 14, pp. 22ff. of which include material derived from interviews like this one.

luminous apparition might be producd by the motion of the fishes in the water, as oftentimes that part of the sea shines, which the prow of a Ship or Boat cutts in its passage, he reply'd, that the sea it selfe did not <then> shine, but the Fishes did so <whether> they were swimming up & down or rested for some time in a place.[52]

Sometimes, what Boyle was told evidently surprised him, stimulating him to a spontaneous follow-up question, as in this instance dated 29 May 1685 involving two planters from New England, probably Nathaniel Weare and William Vaughan:

> The other day, two Gentlemen belonging to the Province of new Hampshire in New England (whence they came not long since) & imploy'd by that Colony to his Majesty here, answer'd me, that in the Winter the Coldest Wind that blowes in their Country, is the Northwest: & being ask'd again, what was their hottest Wind in Summer, they told me it was likewise the Northwest; at which Answer being surpris'd, I ask'd them whether they could give any Reason of so odd a Phænomenon. Whereto they answer'd, that they ascrib'd it to the large Tract of the Continent, & the great Woods that lay to the Northwest, which Woods, they said, in the Winter, had their Branches, through which the Wind past, all laden with Snow: & in the Summer, they said, the close Air of the Valleys, & the thick steams that fill'd it, would conceive so intense a Heat, that sometimes <in the heat of Summer,> when a sudden puff of Wind blew upon their Faces from those sultry vales, it seem'd to them as if it came out of the Mouth of a furnace, & would be ready to overcome them with the Faintness produc'd by the Heat & vapours it brought along with it.[53]

In other cases, there is an interesting element of cross-referencing on Boyle's part, as in the following example involving a French visitor identified as 'Monsieur V.', whom Boyle queried about what he had been told by another informant, João Furtado de Mendonça, who accompanied his father, Afonso, Viscount of Barbacena, to Brazil when he was Governor there in the early 1670s:

> He told that it was true what I had learnd from the Governors son of Brasile, of the great Mountaines of water that sometimes appear on the Coasts when the winds are calme, for he says he observd them particularly 2 or 3 severall times not far from the fort St. Antonio near the great Bay affirming that <thô> the winds were then still yet the waves went so very high, as if they had been near the Cap

---

52   Workdiary 21–699.
53   Workdiary 36–1.

of good hope, so that the fishermen durst not venture abroad to fish for fear of being over set.[54]

The same technique recurred when Boyle interrogated William Bond about Hudson's Bay on 23 March 1687; in this case, he alluded to data about permafrost in Siberia that he had received from his informant on Russian affairs, Dr Samuel Collins, which had been published in the 1683 appendix to *Cold*:[55]

> I spoke to Him of what happen'd in Ciberia, of the Earth that is never thaw'd, if it lye so much as a foot under the surface of the Ground. And He assur'd me that <on> the Coast of Hudsons Bay, in the thick Woods, their shade, & the deep moss that is commonly found at the feet of the Trees, did so keep off or weaken the Sun-beams, that generally at the depth of 6 or 8 Inches, beneath the surface of the Ground it self, they found the Earth hard & frozen, as they had some times occasion to observe to their trouble, when they were to erect some temporary timber buildings in the Woods.
>
> He answer'd me, that He could not say upon his own knowledge, that Hudsons Bay was colder than Russia: yet by some <Swedes & Muscovites> *Yuits*, as he calls them, that had occasion to come to the Bay, He was told that they felt it colder there than in their own respective Countreys.[56]

Sometimes Boyle divulged the hypothesis he was seeking to prove through the questions he asked, in this instance concerning mineral deposits in Cornwall, though his views were almost certainly affected by what he had learned about mines in more exotic locations:

> Another Gentleman an ingenious & diligent observer of Tin Mines assur'd me, that he had not long since observd that in some veins, <belonging to a near relation of his> where the Skillfull Workmen had carefully diggd out the Oar about 8 or 10 years before[,] there was generated in the inside of the Veins, where the air had access, fresh particles of Tin so copious, that they seemd to line the Cavitys made in the Earth; & were judged & found by the Tin-men to be worth the workeing over again; & his Discourse seem'd to confirm much my Conjecture, that a Minerall lying still in the vein where it grew will in a much shorter time affor'd fresh metall in those places where the air has access to it, than if it were diggd out of the Mine before 'twere expos'd to the air.[57]

---

[54]   Workdiary 21–562.
[55]   *Works*, vol. 4, p. 564.
[56]   Workdiary 36–62, 63.
[57]   Workdiary 21–713.

On occasion, Boyle's informant proved disappointing, in this case in the course of a series of questions about coral addressed to an ambiguously identified interlocutor who might again be the Earl of Sandwich, who instead tried to help by volunteering some ancillary information:

> When I demanded whether he had observ'd that any milky Sap ascended to nourish the stony Plant, & whether he had seen any thing like Berries upon it, he ingeniously confessd to me he had not been soe curious as purposely to make enquiry into those particulars, but that he remembred that haveing broken some of the large peices of Corral he took notice that the more internall Substance was much paler then the other, & very whitish, & that at the extream parts of some branches or spriggs he observ'd little blackish Knobbs which he did not then know what to make of. And when I enquir'd what depth the Sea was of in this place, he answer'd that 'twas nine or ten fathom.[58]

Now and again, one senses that, though Boyle recorded what he was told, he may have reserved judgment on the phenomenon he heard about, as in this account of a merman and mermaid, which is perhaps notable for his interlocutor's hasty promise to refer to his notes for fuller details (in fact, though described as 'Mr Wake', with a blank space left for his Christian name, this was probably Nicholas Waite, an East India Company employee who was imprisoned in Manila when on a private trading voyage to the Philippines in the 1670s):

> This day Mr Wake, a person of worth & note, that liv'd some years in the Manilha Islands, told me in the presence of his Brother & another Virtuoso that came with him to visit me, that there was taken upon the Coast there a Merman & a Mermaid, that is a Male & Female of that sort of Creatures, & that He had the good fortune to see them whilst they were yet alive: & when I had ask'd Him some Questions about them He answer'd me, that their Heads were like enough to Human Heads, but their broad & flattish Faces were more like to Monkeys or Baboons. Neither the Male nor the Female had any Beard. Instead of Arms there were fastened to the shoulders certain Organical Parts like great Fins, whose more solid Parts were connected by a kind of strong Membrans; & of those rigid Parts one (if I mistake not the middlemost) reach'd a great way lower than the rest. The stature of these Animals was very great: for He answer'd me that He judg'd the length of the whole to be about twelve foot; & that the length of that Part which seem'd like to a Human Body was not inferiour to an ordinary man. He added, that it was more round than a Human Body. And when I inquir'd after the Compass of it, He promis'd to satisfy me out of the Notes He took in Writing of the Dimensions of the Animal; but said, that as far as He remember'd, the Body

---

[58] Workdiary 21–319. For a comparable example, see entry 536.

was better than 4 foot about. He answer'd me too, that when this Animal was put to swim, the Body lay along in the Water like that of another Fish in the act of Swiming. But that which is the most remarkable in this Creature is, that at some distance below the Navel, it degenerates from Human Shape, & has a Tail like another Fish. The Genital Parts were very manifest, & except that they were not hairy did plainly resemble those of a Man & a Woman.[59]

Much more often, however, we simply have detailed information engagingly delivered, as with the account of 'An odd stinkeing New England Animall' about which Boyle heard from a figure who is almost certainly to be identified as John Talcott, Major of Hertford County in Connecticut:

Enquireing of Major Toms [*sic*] about the Animalls in New England he told me he had severall times observd one peculiar for ought he knows to that Countrey they call it a Skonke 'tis something lesse than an Hare of the Size & Shape of a large Rabbet the Colour Pyebald & the Tail like that of a Fox this Animall makes Burroughs but they are not deep nor much windeing into which if it be pursued by other Animalls, Nature has given it the Defence of a certain Liquor of soe stinking a sent that this diligent Person assurd me that neither he nor any of his Family could for 3 Weekes endure the approach of a Dog that had been perfumed with it.[60]

What should by now be clear is how vivid and intriguing these succinct narratives are. It would be possible to go on giving examples of them almost indefinitely, but the quotations already included should have given a sense of Boyle's earnest and intense questioning, of the types of agendas he was pursuing and of the kind of information this brought to light.

Boyle's Rationale

Moving on from this, it is obviously important to consider how Boyle used the data about exotic regions he thus painstakingly acquired. As has already been noted, it is not coincidental that the interviews proliferate at the point they do in Workdiary 21, because they coincide with Boyle's preparation of some of his most curious books, the series of 'tracts' published in the early 1670s in which he speculated on topics like the nature of sea-water and the saline characteristics associated with it, or the types of organism, such as coral, that flourished on the seabed, or the way in which mineral formations occurred. Tracts like 'Of the Temperature, Of the Submarine Regions, As to Heat and Cold', or 'Relations about the Bottom

---

[59]　Workdiary 36–18.
[60]　Workdiary 21–632.

of the Sea', are littered with verbatim or almost verbatim transcripts of interviews from Workdiary 21, often with the details of the interlocutor exactly reproduced as if Boyle had had such re-use in mind when dictating the notes in the first place (indeed, it is revealing that he sometimes included alternative identifications of the same informant in adjacent entries, as if he wished to avoid repetition).[61] To take a single example almost at random, in his tract on submarine temperatures Boyle exactly quoted from Workdiary 21 the following passage from an interview with Colonel John Vermuyden, a Dutch protégé of Prince Rupert who had sailed up the Gambia river in search of gold in 1661–62:

> Inquiring also of an intelligent Gentleman that was imploy'd to the river of Gambra, & sayl'd up 700 miles in it, in a small frigot, whether he had observed that in the *Sea*, even of those hot climats, wine may be preserved coole, he told me that it might, and that by the means I hinted to him, which was to let down when the ship came to an Anchor in the Evening severall Bottles full of wine (they used that of *Madera*) exactly stoped to ten, 12, or 14 fathoms deep, whence being the next morning drawn up, they found the wine coole and fresh (as if the vessels had been in these parts drawne up out of a well) provided it were presently drunk, for if that Circumstance were omitted, the heat of the Aire on the upper part of the water would quickly warme the Liquor.[62]

In the case of Workdiary 36, its context is slightly different, since we are now in the 1680s, by which time Boyle had reverted to writing full-length books, including such key texts as his *Free Enquiry into the Vulgarly Receiv'd Notion of Nature* (1686). At this point, he was especially interested in medical topics, and not least the extent to which the operation of medicines could be explained in mechanistic terms, the subject of his *Of the Reconcileableness of Specifick Medicines to the Corpuscular Philosophy* (1685). Indeed, this made him particularly interested in poisons and the effects that very small quantities of matter could have on the human body, which may explain why information on such topics commonly appears in these interviews.[63] One account in particular

---

[61] *Works*, vol. 6, pp. 343ff, vol. 7, pp. 413ff. The overlap is only partially documented in the relevant tables in *Works*, vol. 6, pp. xl–xli, and vol. 7, p. xxxv, because the Workdiary edition was still in preparation when the *Works* were published. For the same reason, the identifications of interlocutors offered at http://www.livesandletters.ac.uk/wd/index.html are to be preferred to those in *Works* in cases where the two differ. For cases where the same interlocutor is characterised in more than one way, see, for example, Workdiary 36–47ff. (Bradley), 82–3 (Cony), 91, 94, 97 (Martens).

[62] *Works*, vol. 6, pp. 349–50, exactly following Workdiary 21–675. This follows a typical patchwork of quotations from Workdiary 21–320, 321 and 311.

[63] See Workdiary 36–12, 20, 35, 57, 101. *Specific Medicines* occasionally uses material that sounds as if it derives from workdiaries, for example *Works*, vol. 10, pp. 367, 433. It is

not only deals with poisons and antidotes but also gives a comprehensive account of Indian medicine and of ancillary topics like snake charming, and it is here quoted in full as a further example of the genre. The interview in question was with John Slade, who went to Madras as Master's Mate in the *Surat Merchant* in 1676–77, to Surat as Captain of the *Unicorn* in 1678–79, and to Bantam as Captain of the *Barnardiston* in 1680–82, and it took place on 8 May 1686:

Yesterday being visited by Captain S an observing Gentleman, that had made a good stay in the Kingdom of Bengale, & had sail'd also in the South Sea, & particularly to Bantam; the Answers He return'd to Questions I had time & opportunity to make Him, were to this purpose.

1. That the Indian Brahmanes He convers'd with were very illiterate Men; but yet that the Physicians among them, & other Indians that practis'd Physic in Bengale, cur'd the Diseases of those Countreys happily enough; insomuch that He made use of their Remedyes with good success instead of our European ones.

2. That They seldom imploy'd any Medicines but Simples, & usually but one at a time. And such Remedyes as were capable of the following way of preparation, as Woods, Roots, Barks, & Minerals, (of which last sort they use but very few) they <for the most part> prepard only by grinding them with Water upon a Stone, till they had obtain'd a sufficient quantity of a kind of Mud, of which they gave the Patient the Doses they judg'd fit pretty frequently.

3. That this way they prepar'd a certain Wood, which the Portugese there call *Raeis de Solare* that look'd almost like Juniper & which I found to be bitter in Tast. This Remedy <thô it be not the Bark but the Wood of the Tree> they much esteem against Agues. And Captain S. told me He <had> severall times given it with very good success, instead of Jesuites Bark.

4. That for Dysenteries, which are Diseases very rife in Bengale, their Physicians use a very odd & unpromising Remedy, viz. a strong Decoction <made> of powder'd Rice & black Pepper grossly beaten. This Decoction, being strain'd affords a kind of Gelly, which thô it <tast almost as> strong of the Pepper as if one were chewing that Spice, yet they give it the Patient plentifully. And He added, that He himself try'd it upon sick people with good success.

5. That He knowes an English Gentleman whose Name He told me, that had been troubl'd many Moneths with Spitting of Blood, & great pains in one of His sides, whereby He was brought to be almost a skeleton, & yet was afterwards cur'd by a few doses of a Root that they call in Bengala, *Raeis* (if I mistake not) *Boutoun*.

6. That they have in Bengale excellent Antidotes for the Poysons natural to those Countreys. But that there is one sort of Serpents, whose Name He remembers not, that uses to place himself upon the tops of Trees near Highways

perhaps worth noting that a section on poisons was a further component of the intended content of *Experimenta & Observationes Physicæ*, as noted on p. 185 above.

& Paths, & from thence shoots himself down upon unwary Travellers, & as soon as He has hurt them regains the upper part of the Tree; which sort of Serpents, He sayes, has a Poyson that works so quick & so strongly, that almost alwayes before Antidotes can be apply'd, the Patient dyes. And when I ask'd Him, Whether He had seen any Serpent of this kind, He answer'd me that He never did see any, nor desir'd to do it; the Natives himself [*sic*] being so fearfull of this mischievous Creature, that when they discover it afar off, they fly from it as if it were the Devil.

7. That it might well be credited, which I had heard concerning some Indians, that carry about long serpents coyl'd up in Baskets, & when they please <take> them out, & make them raise themselves upon their Tails, & in that erected posture make divers rude Motions that some call their *Dancing*. For of this Story He among many others had been an Eye-witness. And He answer'd me that thô the Serpents He saw were not so long, yet there were of that kind, that had some three, some four, & some five or more in length. And He further answer'd me, that thô He had not known any Tryal made of the venemosity upon Men, yet having caus'd a Pullet to be bitten by one of them, it dy'd very quickly.

8. That the Stones that are said to be taken out of that sort of Serpent they call *Cobrodi Capelle*, which <sort of> Animals He told me He saw in the Indies, are not really taken in the form we have them, out of the Heads of those or any other Serpents, but are factitious things, & made by the Brahmans with much Ceremony & some superstitious Rites. He further told me that one of the Brahmans inform'd Him, that these Stones were made chiefly of the Ashes of a certain kind of Serpents; but how far they burnt those Animals, <& what they add to make up the Antidotal Stone, and about> other Questions relating to it, I could not be inform'd.

9. That the Antidotal Stones that were genuine or duly made, had really the vertue of curing the bites of venemous Serpents. He added, that to omit other Instances, He was call'd to an English man, that had been bitten in the Hand by a very venemous Serpent, whose Arm He found, before He came, to be very much swell'd & discolour'd; which Symptomes were accompany'd with a terrible Pain, thô there had past but half an hour since He was bitten. But upon the application of the Stone, thô the Tumour persever'd for a good while after, yet the Pain was quickly appeas'd & the danger soon over. of which very Stone the Relator having been pleas'd, <in spite of my reluctancy,> to make me a Present, I found it as big as almost any two that I had formerly either try'd or seen.[64]

Other interviews in Workdiary 36 reflect an interest in atmospheric conditions and related phenomena that had preoccupied Boyle since the time of his initial publications on the nature of the air in the 1660s. His continuing interest in such topics is reflected in his posthumous *General History of the Air*,

---

[64] Workdiary 36–37ff. The following entry, no. 46, gives the weight of the 'great Factitious Snake-stone' with which Slade presented Boyle.

published by John Locke from Boyle's papers in 1692, which contains many nuggets of information derived from the workdiaries.[65] Indeed, we know that at this point Boyle distributed a special questionnaire concerning the air, and many entries in Workdiary 36 have been endorsed with the number of the query in this to which they relate.[66]

The other theme that comes to the fore in Workdiary 36 is the pursuit of data for Boyle's 'Strange Reports', the component of *Experimenta & Observationes Physicæ* that has already been referred to more than once.[67] This workdiary provided four stories for the first part of 'Strange Reports' as published in the volume of *Experimenta & Observationes Physicæ* that actually came out before Boyle's death, dealing with 'things purely Natural'; one of these was an anecdote from Sloane about his visit to Jamaica.[68] As we saw in Chapter 8, the workdiary would evidently also have been one of the sources of the intended sequel to the work, explicitly devoted to 'things supernatural', for which it is clear that Boyle was actively soliciting material in the last years of his life. We saw there how he specifically asked his informants what they could tell him of 'supernaturall' occurrences, and among those who spectacularly obliged was Pierre-Esprit Radisson, whose vivid account of a magical ritual he witnessed among the North American Indians was quoted on page 182 above. Indeed, it is worth pointing out that, although Radisson kept a journal of his own, he there only once showed interest in the kind of matters about which Boyle interrogated him, dismissing these as 'the fabulous beleafe of those poore People', thus again

---

[65] *Works*, vol. 12, pp. 3ff. See, for example, Workdiary 36-1, 27ff., 47–8, 54 and *Works*, vol. 12, pp. 108, 102–3, 105–6. *General History of Air* also includes material from Workdiary 21: see, for example, Workdiary 21–270, 272, 522, 534, 536, 555, 587, and *Works*, vol. 12, pp. 113–14, 137, 148, 42–3, 111, 130. It is noteworthy that many of the endorsements in Workdiary 21 refer to works by Boyle on aspects of the air: see below, pp. 207–8.

[66] On the questionnaire, see above, p. 65 and n. 43. As noted there, the questionnaire as circulated seems to have been differently numbered from the set of 'Titles' Locke included in the published work or which survives in manuscript, as is illustrated by the numbers given by John Clayton in his response (see *Correspondence*, vol. 6, pp. 217ff.); the numbers in the endorsements to Workdiary 36 match those cited by Clayton. This version ranged more widely than that published by Locke, as is reflected, for example, by Workdiary 36–48, 56, both endorsed 'A 25': according to the numbering given by Clayton, this would relate to the query concerning the 'Salubrity & unwholesomenes of the Air' (no. 26 in the list in *Works*, vol. 12, p. xxiii–xxiv), which did not get into print in Locke's recension of the book. Boyle had, of course, published a separate work on 'the Insalubrity and Salubrity of the Air', appended to his *Languid and Unheeded Motion* (1685), ibid., vol. 10, pp. 303ff., in which information from Edward Browne and other travellers is divulged.

[67] See above, pp. 176ff., 184 n. 75, 185.

[68] See *Works*, vol. 11, pp. 434–5, 436–7, and Workdiary 36–21, 85, 86 and 87. On the identification of Sloane, see above, n. 61.

bearing out the extent to which Boyle's earnest interrogations in the workdiaries recorded information that would otherwise be lost.[69]

Yet, even allowing for works Boyle was planning in his final years that failed to come to fruition due to his death, the workdiaries still contain a vast amount of material that he failed to use. Often, the interviews ranged widely, perhaps because Boyle simply allowed his interlocutor to tell him what seemed notable about the places he had visited. For instance, this series of entries in Workdiary 21 ranges from rubies through inflatables to the use of tar in buildings:

> One that was in Cylon confirmed to me <both> that there grow very good Rubyes there as well as in Pegu (where he had alsoe been) and that the Iland produces many Saphyres, of which he had seene severall faire ones, as likewise that in the places where they find the most Rubies, among many of them they diverse times meet also with Saphirs.
>
> The same person answered me that nere Bagdat he had seene the people cross the river upon leatherne baggs (made for the most part of buffe skins) filld with wind and to bee emptyed at pleasure, and that he saw some of them soe bigg that a country man and his wife would goe to markett upon one of them, the man sitting astride on the narrower part, and the woman sitting as on a side saddle on the other,
>
> The same person confirmed to me that they yett build on the parts nere Bagdat not with mortar, but with the Bitumen that is copiously afforded by certaine wells or springs within a few miles of that Citty, which wells he visited.
>
> He alsoe told me that <the> sweete corrosive fruit is ananas.[70]

In another case, Boyle's lengthy interview with a man identified only as 'Monsieur V.' ran through an extraordinary number of topics: an optical illusion due to weather conditions; a strange wind emanating from a cavern; a method of making saltwater drinkable; local variations in the weather; the incidence of diamonds; poisons; methods of mending china and artificially hatching chickens; the nature of the cinnamon and camphor trees and the range of uses to which the coco tree could be put; a unicorn-like animal; the bezoar stone, and the existence of white elephants.[71] Almost every item in this heterogeneous series is endorsed with a note of its subject matter, and it is worth explaining that Boyle frequently made retrospective endorsements to the workdiary entries, which are often revealing in themselves. These commonly itemize the topic

---

[69]   Workdiary 36–102. See also entries 103, 108 and 117. Cf. G.D. Scull (ed.), *Voyages of Peter Esprit Radisson* (Boston, MA, 1885), p. 236, and the account of funeral rites in ibid., pp. 236–40, passim (cf. also p. 36).

[70]   Workdiary 21–514ff.

[71]   Workdiary 21–557ff.

the entry deals with, for instance 'Of the tides not reaching to the Bottom of deep Seas' or 'Observ[ation] about seeming Heat & Cold'; equally often, they indicate the work by Boyle to which he saw them as relevant, as in 'belong to the Aire' or 'To the Deter[minate] Nat[ure] of Efflu[viums]'.[72] But in many cases, no work is specified and the phenomenon in question is simply noted as 'An odd Observ[ation]', as if Boyle considered the nugget of information worthy of record in its own right – for instance, concerning vision in the dark, or an unexpected variation of the compass, or the skin colour of creoles.[73] In one instance, he specifically commented on an 'odd observation for whose sake I chiefly sett down this story'.[74] In many cases, Boyle used some data he elicited from an informant but left part of it unquoted, evidently because it was less directly relevant to the matter in hand.[75] Yet that does not seem to have reduced his sense of the value of the information as a whole, and it was evidently for this reason that Boyle himself seems to have considered it might be worth presenting in its own right, hence bringing us back to the idea of a work entitled 'Physica Peregrinans' with which this chapter began – or, to quote the fuller description given there: 'A pretty large collection of things that I learn'd in conversing with Pilates, Sea-captains, and other Persons that have travail'd unto the Indies, or other remote Country in Order to make Physica peregrinare.'

But what would this actually have comprised? Some items in the workdiaries seem so self-contained that they may well have been intended for onward transmission as they stood, as with the numbered points on Indian medicine stemming from Boyle's interview with John Slade that have already been quoted. In other cases, Boyle may have aspired to a degree of reorganization, perhaps thematic: that this was so is suggested by the first text printed in Appendix 2 to this chapter, in which more than one paragraph that originally started 'He ...' – alluding back to the interlocutor introduced at the outset – was later altered to 'The several times mentiond Priest sent to the King of Siam' or a similar formula, so that each paragraph could stand on its own.[76] Indeed, the groups of re-copied extracts from workdiaries from which the content of Appendix 2 is largely derived evidently themselves bear witness to such a process of reorganization. Here, there is a parallel with another compendium that Boyle planned and partially executed in his later years, his 'Paralipomena': this was intended to supplement his existing writings on various subjects, and it avowedly aspired to a thematic arrangement (indeed, there is some overlap between the intended

---

[72]　Workdiary 21–646, 577, 253 (cf. 256, 260, 350–1, 587, 615–6, 697, 702, 746), 415.

[73]　Workdiary 21–286, 579, 655, 694 and passim.

[74]　Workdiary 21–621.

[75]　In the case itemised in n. 62 above, Workdiary 21–322 and 323, which form part of the same series, are not used. The same is true of the items adjacent to certain of those listed in n. 65 above.

[76]　See Appendix 2, nn. 4, ix.

content of 'Paralipomena' and that of 'Physica Peregrinans', in that a number of interviews with travellers appear in the extant sections of 'Paralipomena').[77] If so, however, the case of 'Paralipomena' suggests that the process of reorganization might have been quite imperfect, with the matter presented remaining rather miscellaneous, recourse instead being had to a complicated system of cross-referencing.[78] As an alternative, the outcome might have been even simpler, perhaps as simple as that exemplified by 'Strange Reports', which presents the texts that it comprises as a series of numbered 'Relations', ten in total for the first, natural part that was published.

What, however, about introductory material? Frustratingly, although Boyle refers at one point to 'A fragment of a tract or Essay call'd Physici Perigrinantes', this no longer survives.[79] The fact that it is described as an essay implies a reflective element which is revealing in itself, and makes it worth considering such clues as exist as to its likely content. These include the comments by Boyle cited earlier in this chapter, particularly the 'Advertisement' to 'Saltness of the Sea', in which he stressed how 'scarce any Writers of Philosophical things' had had opportunities like his to receive 'Authentick Informations' from travellers to exotic lands due to his connections with the East India and Hudson's Bay Companies. Further telling clues are available from Boyle's introductory material to 'Strange Reports'. The published version of this work has only a rather cursory 'Advertisement' which unhelpfully refers to 'what I Wrote about the Nature and Scope of my Collection of Strange Reports, in an Essay which takes its Title from them', another interesting-sounding text which is unfortunately now lost. However, a hint as to its theme is to be found in an alternative version of the advertisement which survives in manuscript and was included in the 1999–2000 edition of *The Works of Robert Boyle*.[80]

As we saw in Chapter 8, both in that and in the published 'Advertisement', Boyle alludes to the work then attributed to Aristotle, *De mirabilibus auscultationibus*, in which, as Boyle explains, 'he threw together without Methodising the collection a pretty number of strange reports, that he had heard concerning divers naturall things', thus providing a slightly unexpected mandate for Boyle to compile a similar collection of his own. But what is significant here is how – after contrasting the greater verisimilitude of the 'particulars' he himself deployed by comparison with his predecessor's – Boyle went on to explain how the retailing of accounts of even questionable phenomena might '<generallie>

---

[77] For an account of 'Paralipomena', including a transcript of various extant portions of text from it, see *Boyle Papers*, ch. 4, passim. For information derived from travellers, see ibid., pp. 206ff.

[78] Ibid., esp. pp. 198–200.

[79] RS MS 186, fols 174v–175. There is a deleted stroke, possibly an opening bracket, before 'or'.

[80] *Works*, vol. 11, pp. lxvii–lxviii, 429. See also above, p. 177.

enlarge the curiosity ‹if not allso the minds› of men & excite those that may
have opportunity to enquire into the truth of these reports ‹in pursuite of›
which enquirys other things may probably be discovered'. This seems crucial, as
does Boyle's further comment that the aim of such anthologies was 'to free the
understanding from being ‹too much› straitend by a confinement to common
things & the phenomena of nature & art familiar to us' and thus 'to suggest
new & more reaching notions about severall things'.[81] The object was to expand
people's conception of what nature could achieve and to make them see that all
sorts of phenomena might exist outside common experience. Moreover, whereas
'Strange Reports' was intended to do this by retailing well-authenticated yet
extraordinary phenomena that were as likely to have occurred in England or
Europe as further afield, the emphasis in 'Physica Peregrinans' was on the sheer
exoticism of the data that illustrated this point. It is worth recalling the title Boyle
gave to Workdiary 21 on its vellum cover, 'The Outlandish Booke', implying that
its subject matter was out of the ordinary, unfamiliar, foreign, remote, bizarre.[82]
It is as if Boyle intended his records of interviews with those who had witnessed
things in far-away places to provide new parameters in natural philosophy: the
claim was surely that such information would challenge people's complacent
sense of the possible, that a worldwide perspective would give a view of the
sheer fecundity of the natural world that would be challenging and exhilarating
in itself.

Boyle's Context

Others shared such an ambition with Boyle. Here, it is appropriate to return to
the early Royal Society, the appetite of whose Fellows for collecting information
about far-away places by means of questionnaires seems to have powerfully
influenced Boyle in the early 1660s, as we saw in Chapter 3. The significance
of such data for the society's conception of the proper agenda for natural
philosophy may be further illustrated by considering the views of its Curator
of Experiments and Cutlerian Lecturer Robert Hooke, who laid repeated stress
on this theme in the Cutlerian Lectures he gave to the society in the 1660s and

---

[81]   Ibid., vol. 11, pp. lxvii–lxviii.

[82]   Cf. *OED*, s.v. 'outlandish'. For a comparable argument for the importance of the
strange and the marvellous in challenging traditional explanations in natural philosophy, see
Lorraine Daston and Katharine Park, *Wonders and the Order of Nature 1150–1750* (New
York, 1998), esp. ch. 6, and Lorraine Daston, 'The Language of Strange Facts in Early Modern
Science', in Timothy Lenoir (ed.), *Inscribing Science: Scientific Texts and the Materiality of
Communication* (Stanford, CA, 1998), pp. 20–38.

thereafter.[83] For instance, in a lecture delivered on 18 December 1697, Hooke stressed the importance of:

> procuring the Description of the Naturall, Geographicall, and Artificiall History of Regions countrys Island[s] seas Citys, towns, Mountains, Deserts, plains, woods and the like places, parts and Regions of the Earth, whether Inhabited and Cultivated, or not Habitable or unfrequented, which may yet afford great Information for the Explication of various phænomena highly necessary to the augmenting and perfecting of Naturall philosophy .... And I doe not at all doubt but that if fitting Means were made use of there might in a short <time> be reaped a plentifull harvest of such substantiall Proper and pertinent materialls as would (when the chaff and tares were separated) be fitt to be put up into the Storehouse or Repository of Usefull <Memorialls or> provisions for the Magazine of Naturall Philosophy.[84]

Hooke also provided a powerful manifesto for the publication of accounts of exotic places in the promotional preface he wrote for Robert Knox's *Historical Relation of the Island Ceylon*, published by the Royal Society's printer under the auspices of the East India Company in 1681.[85] It is no less revealing that a significant component of Hooke's papers comprises manuscript travel narratives or notes on them in his own hand.[86] Though Lisa Jardine has suggested that one set of these – on Jan van Linschoten's *Itinerario* – bears witness to the young Hooke's yearning to travel to such exotic places himself, it seems more likely that, as with Boyle, he was content to collect such data from the experiences of others.[87]

A further example is provided by John Locke, another avid reader of travel books who clearly saw the data they provided as integral to natural philosophy. Such works comprised one of the largest components of Locke's library, and they provided him with illustrative material that he used in his writings, not least his famous *Essay concerning Human Understanding*, which was heavily indebted

---

[83] See Hunter, *Establishing the New Science*, esp. pp. 301–3 and 332–4.

[84] Royal Society Classified Papers 20, 80, partly quoted in Jardine, *Curious Life of Hooke*, pp. 285–6. Within the quotation, 'and' has been deleted before 'Proper' and after 'chaff', and 'R' has been deleted before 'Storehouse'.

[85] Robert Knox, *An Historical Relation Of the Island Ceylon* (London, 1681), sigs (a)2–4. For a commentary, see Joan-Pau Rubiés, 'Instructions for Travellers: Teaching the Eye to See', *History and Anthropology*, 9 (1996), 139–90, on pp. 139–41.

[86] See esp. City of London Guildhall MS 1757, passim, but also Trinity College, Cambridge, O.11a.1, items 10 and 57, and a few items in British Library MS Sloane 1039, for example fols 133, 135–6, 139.

[87] Jardine, *Curious Life of Hooke*, pp. 82–5 and 344.

to the literature of travel.[88] Locke also solicited information about phenomena in exotic locations from his correspondents, even communicating information 'about poisonous Fish in one of the Bahama Islands' that he obtained from one of these to *Philosophical Transactions* in 1675.[89] Locke's interest in the data Boyle gleaned from his overseas informants is clear not only from his inclusion of many such accounts in his edition of *The General History of the Air*, but also from the fact that he had a fair copy made of the accompanying data that Boyle obtained about tarantulas, as reproduced in Appendix 2 to this chapter.[90] Indeed, Locke occasionally recorded in his journal similar conversations he had himself, for instance with the French traveller and natural philosopher François Bernier in Paris in 1677–79, thus providing a limited echo of the massive activity on Boyle's part that has been surveyed here.[91] Yet the contrast in scale is revealing in itself, and it is worth noting that, though Hooke's and Locke's interest in the testimony of travellers provides a significant context for Boyle, neither of them offers any real analogy to the extraordinary record to which this chapter has been devoted. Boyle's ability to summon so many distinguished or expert visitors to divulge their knowledge to him in oral form seems to have been unique: the combination of his aristocratic status and the renown of his publications on natural philosophy meant that visitors from England and abroad were prepared to pay court to him in the way that would not have been the case with lesser men like Locke and Hooke.

A further point concerning Locke's use of travel literature has been made by Daniel Carey. In *Some Thoughts concerning Education*, Locke commended such writings as containing 'a very good mixture of delight and usefulness', and Carey has observed how 'knowing that Locke found this material pleasurable confirms the sense we have that it fueled his imagination and produced a particular kind of delight'. Indeed, Carey has suggested that travel books may

---

[88]   See Peter Anstey, *John Locke and Natural Philosophy* (Oxford, 2011), pp. 59–61; Daniel Carey, 'Travel, Geography and the Problem of Belief: Locke as a Reader of Travel Literature', in Julia Rudolph (ed.), *History and Nation* (Lewisburg, PA, 2006), pp. 97–136, and Ann Talbot, *The Great Ocean of Knowledge: The Influence of Travel Literature on the Work of John Locke* (Leiden, 2010).

[89]   Anstey, *John Locke*, pp. 60, 62. See also Daniel Carey, *Locke, Shaftesbury and Hutcheson: Contesting Diversity in the Enlightenment and Beyond* (Cambridge, 2006), ch. 1.

[90]   See below, p. 227.

[91]   See John Lough (ed.), *Locke's Travels in France 1675–9, As related in his Journals, Correspondence and other Papers* (Cambridge, 1953), pp. 177, 200, 282. For comparable diary entries by Hooke based on conversations with Robert Knox, see Jardine, *Curious Life of Robert Hooke*, pp. 280–81; see also, for example, the details of a conversation with Sir John Narborough recorded in H.W. Robinson and Walter Adams (eds), *The Diary of Robert Hooke 1672–80* (London, 1935), p. 271, though in general Hooke's diaries are too telegraphic to record comparable data.

have offered Locke pleasure comparable to that which he had gained in his student days from romances, offering 'an imaginative outlet to Locke while complementing his disciplined and serious intellectual pursuits'.[92] There is a significant analogy here to the case of Boyle, who had been a great devotee of French romances in his youth: as we have seen, his earliest workdiaries comprise lengthy extracts from such works, while his first literary efforts were heavily influenced by writings of this kind, including his retelling of the story of the early Christian martyr Theodora.[93] Later in his life, we know of only one instance when Boyle reverted to the romance form, in a work called 'The Aspireing Naturalist (a Philosophical Romance)', described in a list of his writings as 'containing an accompt of some inventions & practises said to bee in use among the Inhabitants of an Island in amity with the new Atlantis'.[94] It is surely significant that, in Boyle's case as in Locke's, this residual penchant for the romance form seems to have been related to the interest in far-away places reflected by his reading of travel books and his interviews with travellers to exotic locations.

Only a fragment of 'The Aspireing Naturalist' now survives, which is printed as Appendix 1 to this chapter, but this suggests that the work was intended to illustrate themes similar to those that formed the probable rationale of 'Physica Peregrinans'. The allusion to Bacon's *New Atlantis* in the description of the piece is revealing in itself, providing further evidence of Bacon's profound influence on Boyle, and the objective of 'The Aspireing Naturalist' was similar to that of *New Atlantis* – namely, to show how normal expectations of what 'is possible to be performed by Medicine, or by Art' could be transformed by 'the power of Art & Nature conspireing', particularly in relation to human health. This was to be exemplified by the inhabitants of a far-off island, whose products infinitely exceeded estimates based on readers' 'scant measures of what is wont to happen, or be done, in their own Country; or by the more familiar Phænomena & common operations of Nature', yet which would not be surprising to those of 'a free and empancipated Genius' whose conceptions had been liberated by such knowledge.[95] This exactly

---

[92]　Carey, 'Travel, Geography and the Problem of Belief', p. 114.

[93]　See *Boyle by Himself and His Friends*, pp. xv–xxi; *Between God and Science*, ch. 4; and *Works*, vol. 13, pp. 3ff. See also Lawrence M. Principe, 'Virtuous Romance and Romantic Virtuoso: The Shaping of Robert Boyle's Literary Style', *Journal of the History of Ideas*, 56 (1995), 377–97, and above, pp. 35–6.

[94]　*Works*, vol. 14, p. 333, Boyle's list of his 'Tracts' dated 19 November 1667. There is also a summary of the thrust of the work in *Occasional Reflections*, *Works*, vol. 5, pp. 171–2, though it is interesting that this makes it sound as if it would have had more of a focus on manners and customs (as with Locke's predominant use of such material), in contrast with the stress on 'the power of Art & Nature conspireing' in the extant fragment.

[95]　See Appendix 1.

echoes what has been suggested here about the role of the exotic in 'Physica Peregrinans' on the basis of Boyle's prefatory comments to 'Strange Reports'.

Despite such hints, the extant fragment of 'The Aspireing Naturalist' is rather opaque and convoluted, its meaning at times being almost drowned under the burden of its complex syntax. In this, it resembles the rather wordy rewriting of the second part of his *Theodora* that Boyle executed in his later years, in contrast to the direct and powerful version he had produced in his adolescence.[96] Yet, to return to 'Physica Peregrinans', what is extraordinary about the narratives from the workdiaries that the work would evidently have comprised is how vivid and arresting these are. It is if, by dictating the details of his interviews to his amanuenses soon after they occurred, Boyle achieved a directness and spontaneity that all too often eluded him in his finished writings in his later years. The conception of an anthology of such material may suggest that he himself sensed this, and that he hoped that the striking narratives about strange places and phenomena he had recorded might not only encourage people to expand their conception of the true scope of nature, but might also prove entertaining in their own right. Perhaps he wanted to make a novel contribution to the travel literature of which he was so avid a reader, hoping that the telling details his interviews brought to light might prove engrossing in themselves and thus show that natural philosophy could be diverting as well as instructive. Indeed, on the basis of this material, Boyle may perhaps claim a more significant role in the development of the burgeoning taste for travel literature in the late seventeenth and eighteenth centuries than has been realised.[97]

The point may be illustrated by one final example of an interview, in this case probably with John Harrington, an employee of the East India Company on the Malabar coast who was captured during the Second Dutch War and incarcerated first in Colombo and then in Batavia, whence he escaped in 1667.[98] This lengthy narrative, replete with its own title, illustrates all the facets of these interviews as outlined in the course of this chapter – the vivid

---

[96] See *Works*, vol. 11, pp. xi–xiv and 3ff.; vol. 13, pp. xxi–xxiii and 3ff.

[97] See, for instance, P.J. Marshall and Glyndwr Williams, *The Great Map of Mankind* (London, 1982), ch. 2; Percy G. Adams, *Travel Literature and the Evolution of the Novel* (Lexington, KY, 1983), ch. 2; Michael McKeon, *The Origins of the English Novel, 1600–1740* (Baltimore, MD, 1987), pp. 100ff; Neil Rennie, *Far-fetched Facts: The Literature of Travel and the Idea of the South Seas* (Oxford, 1995), esp. ch. 3. For the rather different claim that Swift derived the idea of *Gulliver's Travels* from the reference to Boyle's romance in *Occasional Reflections*, see William Oldys and Joseph Towers (eds), *Biographia Britannica* (6 vols, London, 1747–66), vol. 2, p. 920n.; see also above, n. 94, below, p. 218, and Principe, 'Virtuous Romance and Romantic Virtuoso', pp. 390–91n.

[98] Workdiary 21–722ff. The inserted passages in paras 7 and 10 comprise passages of text continued in the margin.

details Boyle brought to light, his insistent questioning and cross-referencing, and the range of topics that interested him, from the temperature and terrain of the seabed to the experiences of the divers themselves. As should by now be clear, it is addictive stuff, and it is to be hoped that readers will have recourse to the online workdiaries in search of further examples for themselves:

> Observations about Divers obtain'd by
> Questions propos'd to an inquisitive
> Travailer who was present
> at the famous Pearle-
> Fishing at Manar
> between the Iland
> Of Ceylon & the
> Neighbouring
> Continent.

Hee told me that the Divers (who were allmost all Negros not Indians) though they went downe with a rope ti'd to each of them to be drawne up againe by, yet they did not sinck themselves by weights, but swamme to the bottom with their heads downwards & that this solemne pearle Fishing is usually perform'd about the month of August.

Hee told me that in some places they Dive about 30 braccia (as they call the measure about 5 foot long;) in some places, 40 50 or 60. & in some 150 or 200 braccia: but in some places they were not able by their fathoming lines to reach any bottom.

Hee told me that they oftn met divers great Fishes, in the depths but were seldom much harm'd by any but those they call Tuberones (perhaps our Sharks), that would sometimes bite off an Arme or a Leg or a great gobbet of flesh.

That the deeper the places were that afforded the pearles, the better qualify'd (for the most part) those jewels were; nor were the Divers permitted by the King's Officers, though yet they would doe it by stealth, to seeke for pearles beyond such a depth, beyond which the Fishing was reserv'd for the King.

That the Fish wherein they found the pearles is a kind of Oyster (but larger than ours) whereof the Relatour did eate many. they lay them in heaps to op'n & putrifye: & then takeing out the Pulpy part, they crush & rubb it between their hands to find out the pearls.

Hee told me that the Divers did not, ev'n at the greatest depths, complaine of the weight of the water above them.

That Hee observ'd some of them by his watch to continue under the water ¼ of an houre & many of them to endure halfe that time or some of them us'd only a kind of whistle, but very <much larger than an ordinary one to help them to some respiration, but others us'd nothing at all.>

That the Divers when they were at the bottom of the sea at a great depth could not see before them near ½ yard, but were environ'd with much darknesse, which reduc'd them to find out the Fish by groaping not by sight.

When I ask'd him whether they found not the bottom of the sea very unev'n, hee answer'd that they did, insomuch that, the soile being for the most part very stony, they found divers rocks (some of them worthy to be call'd litle Hills) some very steepe & others that seeme to consist of many vast stones with store of Sand about them. as some of the divers found by groaping along for Fish.

When I ask'd him whether they did not find it cold at the bottome of the Sea, hee reply'd that they complain'd very much of the great cold they felt there, especially if they had gone very deep; & that hee saw divers of them come shivering out of the water with soe much sense of cold that notwithstanding the heat of the climate, there was usually a fire kept to warme themselves at before they were to Dive againe. And some <of those told him that when they had occasion to dive deepe neare the disemboging of rivers they found it much colder than in those parts of the Sea that were remote from fresh water.>

When I ask'd him whether the sea water were not salter at the bottom than at the top he answer'd it was, <&> that not only the Divers found it soe by the tast but, that they oft'n times brought up with the oysters certaine stony Lumps to which the Fishes were joyn'd, which were very salt (as he found by the tast); being divers of them cover'd with a thick crust of somewhat a darker colour than French Bay Salt, & soe strongly saline that severall of the Fishermen & other poore people imploy'd it to salt their Fish with.

When I ask'd him whether the Divers could at the bottom of the Sea perceive any operation of the rough winds that blow'd at the top or of the Currents; He told me that as to the winds they cold not, but the motion of the Currents were sometimes soe sensible at a great depth under water, that the Divers were faine to take hold of great stones or shelter themselves among them to avoid being carry'd away or very much disturb'd in the worke of the Currents.

When I ask'd him whether it were not very easy to draw up the Divers from the bottom, he reply'd that 'twas soe very easy that sometimes for curiosity sake at one pull of the rope [he] rais'd a Diver that was near the bottom, about 4 or 5 Fathom (as he found by the laxity of the remaineing rope[)].

Hee answerd me also that 'twas true that at the Fort of Batavia (where he had been a Prisoner) most of the Stones (some of which were very exceeding large) had by reason of the penury of such materialls on the shore been fetch [*sic*] from the bottom of the Sea by the help of Divers, who there fastn'd them to the ropes by which they were drawn up.

On this occasion I remember that Speakeing with one that liv'd at Goah, I ask'd him whether 'twere true that Fresh water was fetchd thence from the bottom of the Sea by Divers; to which he answer'd that he had severall times been present when the Divers brought up fresh water near Goah, but that it did not come (as some have written) from a fresh Spring riseing at the bottom of the Sea, but, according to his opinion, from a river of fresh water that runs impetuously enough into the sea near the Iland, & by two wheeling Tides is cover'd with Salt water, whence he observd that near the surface the Liquor was very Salt & afterwards lesse & lesse brackish; & when the Divers came to a considerable depth beneath the surface the water would be found Fresh.

Appendix 1
Extant Portion of Boyle's 'The Aspireing Naturalist (a Philosophical Romance)'

Two copies survive of the opening paragraphs of Boyle's 'The Aspireing Naturalist (A Philosophical Romance)', both in the hand of Robin Bacon, Boyle's principal amanuensis in his later years; the title appears in the top left-hand margin of the first page of each. The more continuous version appears in Boyle Papers 9, fols 43–4, where a catchword makes it clear that the text on fol. 44 is intended to follow that on fol. 43 (in the title in this version, 'A Philosophicall Romance' is not enclosed in brackets). The second copy is to be found in Boyle Papers 8, fols 206–7 (containing the equivalent text to BP 9, fol. 43; the lower part of the second page of the bifoliate is blank) and 210 (containing the equivalent text to BP 9, fol. 44). In contrast to the BP 9 version, this has a few alterations and additions: one of these is in the hand of Robin Bacon, and is incorporated into the fair copy in BP 9, but the others are in the hand of another favoured amanuensis, Hugh Greg, and these make additions to the text which do not appear in BP 9 version. The version in BP 8 has therefore here been adopted as the copy text, with the substantive differences between it and that in BP 9 noted, though minor differences of orthography have been ignored (as have minor orthographic adjustments to the text in BP 8 itself – for example, the alteration of 'doe' to 'do').

In addition to the description of the work in a list of Boyle's 'tracts' dated 18 November 1667 quoted on page 213 above, the work is also alluded to in his *Occasional Reflections* (1665), which explains how he 'had thoughts of making a short Romantick story, where the Scene should be laid in some Island of the Southern Ocean'.[1] A further passing reference to 'the account of our Island' appears in entry 802 in Workdiary 28, a compilation of ideas for works by Boyle dating from the early 1670s. Of the interlocutors, *Philaretus*, 'lover of virtue', alludes to the name Boyle adopted for himself in his autobiography (the name is also used in certain dialogue-like sections of *Occasional Reflections*), while *Authades*, 'self-opinionated', is the name of an interlocutor in the appendix to Part I of *The Christian Virtuoso*, which takes dialogue form, though he is referred to rather than speaking himself.[2]

---

[1]   *Works*, vol. 5, pp. 171–2.
[2]   Ibid., vol. 5, p. 95; vol. 12, p. 370.

'A Fragment of *The Aspireing Naturalist* (a Philosophical Romance)'

Having spent some time in Entertainments of this nature, which were proper to make way for the Discourse that was to follow; *Authades* being not forgetfull of his promise, and judging by my lookes, that it was in my thoughts, and that I was perhaps about to mind him of it, began to pause a while; either to recollect, or marshall the things he had to say, or else to prepare an attention, which my respect for him, and the importance of what he was to treat of, had sufficiently bespoken; and afterwards turning his Eyes on me, he began with a somewhat smileing countenance to speake after this manner.

If I looked upon you, *Philaretus*, as a Person that either had a soul straiten'd by the narrow Opinions & Prejudices of the vulgar, ev'n of Learned men, or that were unacquainted with any thing in Art or Nature, save what is common, I should repent that I let fall any words capable of the interpretation (which I perceive by your silence & looks you gave[i] them) of a promise, that you expect I should now make good unto you. For those that are not of a free and emancipated Genius, and are not accustomed to see, and to make Reflections on, the wonders of Nature and of Art, cannot but be inclin'd to estimate the possibility of them both, by the scant measures of what is wont to happen, or be done, in their own Country; or by the more familiar Phænomena & common operations of Nature. Whereas, from the power of Art & Nature conspireing, they will expect most, that have been most conversant with them; that power being indeed such that scarce any will entertaine apprehensions not derogateing from /fol. 207/ it, that have not had some considerable experience of it. But the acquaintance I have with you[ii] inviteing me to suppose you a qualifyed Auditor of what I am about to say, though I well know that very many of your Learned men in Europe find the seeming disbeliefe of all strange things, an easy way to be believ'd judicious; yet I shall not scruple to advertise you, that you are not to thinke that I say not Nature & Art, but the Philosophers of our Island, are able to performe no other, nor no greater, things then those I am about to name to you; since Discretion confines me to mention only those, which by the things that some of you in Europe know already, I may keep you from thinking too extravagant and incredible. /fol. 210/

That the things I have to mention may appeare the lesse <uncapable of your Beliefe>,[iii] it will not be amiss to premise two or three preparatory considerations.

And first you may be pleased to take notice, that I enter upon this Discourse with a very great disadvantage; for the Principall things I am to acquaint you with being Matters of Fact, and consequently such as would be the most properly prov'd out of the Bookes of Observations of our Naturalists & Physitians, and out of other Historicall writeings (wherein our men think it a great impiety not to be scrupulously faithfull, and would find it a great disparagement not to be very punctuall) you are utterly a stranger to these Relations; by which meanes I foresee I shall be reduc'd to employ, instead of these that would better accommodate me, the observations, and other Historicall Narratives, of your European Physitians and other writers; which therefore it will be but just that (as to those that are deliver'd by men of credit) you allow me to make use of, and to argue upon what they affirme, as to what is possible to be performed by Medicine, or by Art. For, though in what I relate especially as to strange Diseases & Recovery's, all is wont to be ascrib'd chiefly to nature, or partly rather to chance; yet *we* judge it warrantable enough to thinke, that what nature may do, when guided but by chance, she may better performe, when directed by Art & skill; whereby she may not only <be> regulated <in her unusual sallys: & odd motions,> but, where it is needfull, be <brought to [rule]>.iv We see that as some soporiferous Diseases, such as one sort of Coma & the Lethargy, are excited in an humane Body, by a peculiarly temper'd morbifick matter; so the like Torpid Affections may be produc'd by Opium; and there are certaine salts, that, as they are very usefull Medicines in those Diseases, so do they likewise powerfully correct the Narcotick quality of Opium. We see also that the tickling of the Nostrills with a Straw, or Feather, may as well cause sneazing as a cold in the head, or the shineing of the sun upon it.

Textual Endnotes

i   Altered from 'give' in BP 8, fol. 206. BP 9 has 'give'.

ii   Altered from 'your'.

iii   Replacing 'incredible to you' deleted. The revised version appears in BP 9, fol. 44.

iv   Inserted in BP 8 with the intention of replacing 'over-rule them', though that was not deleted; 'rule' is accidentally omitted from the insertion, and has here been included for the sense. BP 9 here has 'over-ruled' (it also has 'be' before 'regulated' in the previous line, as inserted in BP 8, but not the longer inserted phrase).

Appendix 2

'Physica Peregrinans, or the Travelling Naturalist': Narratives from Hitherto Unpublished Manuscripts

As has already been noted, the content of 'Physica Peregrinans' would almost certainly predominantly have derived from the vivid and absorbing narratives to be found among Boyle's workdiaries, particularly Workdiaries 21 and 36. It is hoped that readers' appetites will have been whetted by the profuse quotations from these sources included in the text of Chapter 9, and that they will be inspired to go to www.livesandletters.ac.uk/wd/index.html and browse this immense body of material for themselves. Here, it seems worth taking the opportunity to supplement the larger body of material available online by publishing for the first time some examples of the records that Boyle kept of conversations with visitors to exotic locations to which Chapter 9 was devoted; it seems almost certain that these, too, were destined for 'Physica Peregrinans'. They survive scattered through the Boyle Papers, mostly falling into the category of material transcribed from workdiaries that are now lost which has been surveyed elsewhere.[1] All are in the hand of Robin Bacon, except for two – the one about the tarantula, which is in the hand of Thomas Smith, another amanuensis whom Boyle used extensively in his later years, and the very first of all: this is mostly in 'hand C', another hand found frequently in documents written in the 1680s, the author of which has not been identified by name, while the final paragraph is in the hand of Hugh Greg.[2]

As noted in Chapter 9, these appealing records raise all sorts of questions about Boyle's motives in conducting his interviews and the conclusions that can be drawn from them. Apart from anything else, the sections presented here well illustrate the direct, uncomplicated prose in which the accounts are written, and their vibrant, everyday phraseology, even including words like 'cow-turd'. We encounter a typical range of the sort of figures whom Boyle interviewed, ranging from Antoine Pascot, a French missionary to the Far East, and Sir Thomas Rolt, an official of the East India Company, to an unidentified employee of the Hudson's Bay Company and more miscellaneous 'Travellers' and others, identified by

---

[1] See *Boyle Papers*, pp. 167–76, including a calendar of material of this kind.
[2] For details, see ibid., pp. 47–9, 53.

Boyle's typical circumlocutions. The interviews also illustrate Boyle's insistent questioning, including the way in which he indulged in follow-up questions during the course of an interview. Other themes discussed in Chapter 9 that recur here include Boyle's tactful statement of disappointment when those in the field failed to pursue matters as fully he would have liked, as seen particularly with the 'good Fathers' who lacked the equipment to dissect the two-headed snake they encountered in Siam. The interview with Pascot concerning Siam is also revealing for the way in which the account was edited to include multiple identifications of its interlocutor so that its components could be redistributed thematically. The Hudson's Bay text is notable for the pencil endorsements to each of its paragraphs in Boyle's own hand, illustrating the points that seemed to him significant and his tabulation of them. Above all, the information Boyle obtained is fascinating and often extraordinary, perhaps reaching a climax with the use of the philosophical mercury in conjunction with herbs in the East Indies as recounted by Sir Thomas Rolt.

A French Missionary from Siam[3]

A secular Preist instructed in the Cartesian Philosophy by Monsieur Rohaut and imploy'd into Europe by the King of Siam haveing favour'd me with two or three of his Visits answer'd Juditiously several questions that I ask'd him.[4]

---

[3] The document is endorsed in the margin near the start 'Tb'd' in the hand of Bacon. A copy in Bacon's hand of the first few lines of this narrative (ending abruptly at 'well-colourd' in the third sentence) survives in BP 35, fol. 204; the text is identical apart from minor differences of orthography except at two points: in l. 1 'into' there appears as 'in' and in the second sentence 'cheefly' is there mistranscribed as 'clearly'.

[4] Boyle's informant was Antoine Pascot (*c.* 1646–1689), a French missionary to Siam who, with another missionary, Bénigne Vachet (1641–1720), accompanied an embassy sent by the King of Siam to France in 1684. From there, the ambassador and Pascot visited England. The status of Boyle's informant is further clarified by a marginal note added in Bacon's hand towards the bottom of fol. 76, marked to replace 'He' at the start of the third paragraph, evidently as a connecting passage enabling the paragraph about snake stones to be quoted separately from those that precede it here (see above, p. 208): 'The Secular Priest I elsewhere mention'd <who> was imployd by the King of France to him of Siam, & is a Person sober & vers'd in the New Philosophy being askt by me what I might think of the vertues ascrib'd to the Snake Stone brought as from the East Indys, in which <countrys> he had travel'd & dwelt many years; This Person I say ...' ('Tb'd' has been written in the margin at this point, and again adjacent to the deleted 'He' in the text; within the passage quoted here, a character has been deleted before 'ask't', and 'dwelt' was replaced by 'liv'd' but marked for reinstatement.) Boyle refers to this meeting and to the first matter raised in it in Workdiary 36–79. His reference in the text to the eminent Cartesian Jacques Rohault (*c.* 1618–1672), whose lectures at his house at Paris were famous, confirms that his informant was Pascot,

The Phylosophycal particulars[i] that I learned by our Conference were cheefly these. He answer'd me that in a River of Cochin china where Native Gold is often taken up there was found & brought to the King as I remember for a great Rarity one Graine or lump of Native <gold> that weighed a pound & five ounces[:] this he saw & told me that it was well colour'd & seem'd to be meer Gold but the surface of it was very Irregularly figur'd.

He answer'd me that the Tides about the Capital City of Siam are extreamly Irregular in so much that a Captaine of a Great Dutch ship that had stay'd there about three year tho an Ingenious man & an Inquisitive observer of these Anomalies confessed to him he could make nothing of them but my relater staying with some of Disiples[ii] eleven months in a place nearer by many Leagues to the Mouth of the River & makeing frequent & heedfull observations sometimes himselfe & sometimes by his Dissiples of the strange Phænomena of the Tides he found them <Less> Irregular then they were higher up the River but yet that there was in them a wonderfull Irregularity in comparison of the Tides of Europe & also other parts of the East Indies notwithstanding haveing dilligently survey'd the Cituation of the Coast the windings of the River the Islands that are near its mouth & the[iii] Monsons & other winds he thought he could give a Tollerable account of these surprizeing Phænomena.

He answer'd me that that there was indeed Great Vertue in the stone that is said to be found in an Indian serpents head provided it be genuine. He said that he applyed it indifferently to the biteings & stings of all the /fol. 76v/ Venomous creatures in those parts. He said his way of useing it was with the[iv] point of a penknife to scarify <a little> round about the <wounded> place & then aply the stone puting <round> about the Edges of it some Venice Treacle & at the same time giveing the patient about the bigness of a Larg pea of the same antidote & he affirmd to me that he had cured by this means above six score several persons & <al>most[v] all of them with the selfe same stone: its vertue inviteing him on all occasions rather to make use of that than any other.

He affirm'd to me & others that he had really seen & handled a serpent with two heads at the two extreams of his body, so that he could go forwards & backwards without turning his whole body as other serpents must do. For the sake of one that was bitten[vi] he sought dilligently for the animal that had done the mischief & haveing found it he & his company seas'd upon it & where as he had before seen another of the same kind but Transiently he brought this to some Learned men that were Missionarys & veiw'd it at Leasure & they all concluded that it had two <real & true> heads[;][vii] it was about the

---

who had a philosophical background that Vachet lacked. See Adrien Launay, *Mémorial de la Société des Missions-Étrangères* (2 vols, Paris, 1912–16), vol. 1, pp. 6–7, 10–11, vol. 2, pp. 493, 614–15; John Anderson, *English Intercourse with Siam in the Seventeenth Century* (London, 1890), p. 243.

bigness of a viper & about 10 Inches in length[.] I would faine have knowne somewhat of the structure of the Internal parts of this strange Animal but these good Fathers either had not the curiosity <& skil> or had not at hand fitt Instruments to make the Desection. <This>[viii] kind of serpents is very venomous yet the bite of that which he Took was cur'd by the applycation of the above mention'd stone. I remember a Greek Native of Morea assured me that he had seen several Amphis Bæna's in that Country.

He[ix] answer'd me, that on the shores of *Cochin-china* & no<where> else in all that Indian Coast, there is a sort of wonderfull sea fish which resembleth much our Crab [& seems to me to be a peculiar species of them] which go neither streight forwards nor directly backwards, but side wayes, & that with such Celerity, that one may often see them through the /fol. 77/ waters run on the strand: those being taken out of the sea quickly loose their Lives, & soon after begin to loose their Consistence, & within about a week (for I inquir'd after the time) they grow to be of a stony hardness. He told me he had one of them in his hands, that having been lately taken out of the sea was yet soft. He sayes They are much prized by the Natives for their Medicinal vertues, the harder of them being given with great success in Feavers & some other Diseases. He brought over, as he told me,[x] as a Present for a great Minister of State, a couple which were taken in the act of Generation, & being Petrify'd retain their former union or connexion. But these were closed up in a Boxe which he durst not open. Two others but single he brought into Europe, one of which he was pleas'd to present me. The Legs are lost, but the Body is entire, being hard & ponderous. And the shells belonging to Crabs are very visible on the surface. But I was loth to make any Tryals upon this Present, for fear of spoyling so great a Rarity.[5]

[BP 39, fols 76–7]

Hudson's Bay[6]

An industrious Person employ'd by the company of Hudsons Bay, and who lived several years on the coast that embraces it, being come to me this day about an Affair of the company's, I ask'd him some questions, to which he answer'd me.

---

[5]   Boyle's attitude differs here from that displayed in the instance noted on pp. 139–40 above. In the first sentence of this paragraph, the square brackets are Boyle's.

[6]   This series of extracts, evidently copied from an otherwise lost workdiary, probably dates from the 1680s. The identity of Boyle's informant is unclear. The text is in Bacon's hand, but the inserted word in the fifth paragraph is in the hand of Greg. Throughout, endorsements indicating the content of each paragraph have been added in the margin in pencil, evidently in Boyle's own hand; these are here presented in footnotes. They are occasionally hard to read.

That, thô in some parts of that Bay the cold be greater then in Norway, yet where the Ice was uniform, and made by congelation of quiet water, not by accumulation of Flakes of Ice sliding over or under one another, he never knew the thicknes of the Ice to exceed four foot; and usually when 'twas bored and measur'd, it was but three foot or a little more.[7]

That in places shelter'd from the wind, as particularly in some woods, where the snow could not be accumulated, the depth of it seldom exceeded four foot.[8]

That in very sharp winters, the earth or ground was not frozen above two foot and a half, or at most three foot deep.[9]

That the Frosts in Hudsons Bay last about six months of the year, beginning about the 20th of October, and continuing till about the 10th or 20th of April; during all which time there seldom happens any Thaw that lasts above some hours. And, that the Thaw is not wont to be compleated till September; and even <then> it do's not usually reach so far, but that, when men have dug a foot and a half or two foot below the surface of the earth, they find underneath a pretty thick Bed or Layer of ground, that remaines still frozen.[10]

That the coldest winds, and the most boisterous, that blow in those parts come from the north west: the easterly and north easterly coming over a lesser Tract of Land, being also lesse cold: and[xi] the southerly wind, that blows over a great Tract of Sea to arrive at the Bay, being the least cold of all.[11]

That the Summers are very hot, so as to be troublesome on that score, and during that season they are much infested with Musketo's; but yet if at any time during those heats, the north west wind chances to blow, they feel within a few hours, a /fol. 50/ cold almost like that of the winter, and are fain to betake themselves to their warm cloaths again.[12]

But all this coldnes is not the effect of that wind, barely as 'tis wind, that is a stream of Air; because they feel a notable coldnes in the Air, whilst yet it continues calm; and there are yet some hours to come before the Arrival of the Wind.[13]

The same Person likewise told me further, that one[xii] perceiving an unmovable piece of Ice, of a stupendious bignes & hight, and many Leagues off the Shore; He and his company made up to it, and found it to be a ground, as appear'd by a strong current that ran along the sides of it, with divers Masses of Ice floating on it.[14]

---

[7] In margin: 'Thickness of Ice at HB'.
[8] In margin: 'depth of snow'.
[9] In margin: 'How far the Frost goes'.
[10] In margin: 'Duration of Frosts' and 'Odd Observation about Frost'.
[11] In margin: 'Obs. about Coldnesse of the Winds'.
[12] In margin: 'Hot sumers s ... changes'
[13] In margin: 'NB'.
[14] In margin: 'Prodigious thickness of Ice'.

Wherefore out of curiosity they sounded, and were surpriz'd to find, that it reach'd beneath the surface of the sea at least 190 Fathome. I say *at least*, because at that depth they could not fetch ground, and their sounding Line was no longer /fol. 51/

The same Person answer'd me, that he had seen one of those Whales that affords the *Sperma Ceti*; and that they get some of it, thô not an equal quantity, from the other Parts than the Head; and that they obtain'd of it more or less from the Blubber[xiii] or Fat, in what part ever of the Fish they found it. This sort of Whales, He sayes, is very scarce.[15]

He likewise assur'd me that he had seen in those seas several Fishes, that bear such Horns as commonly pass for Unicorns; and remembers one Fish particularly that put his Head out of the water, so that his Horn, which was of a considerable length, might be plainly seen.[16]

The Shape of the Fish, he could not well discern, but the bignes he guess'd to be near double that of an Oxe.

He also answer'd me, that he had met with white Bears 40 Leagues out at Sea; and that he had seen one of that sort of Animals, which thô, by reason of its shape, it did not appear so much bigger than a large Smithfield Oxe, (for that was his Phrase), yet he judg'd it to exceed it very much in weight: for having cut off one of its Paws, they found That (without the Leg) to weigh above 50 lb. but for the length of the Beast, having omitted to measure it, He could not tell me, thô my question was grounded upon the Relation I mention'd to Him, of some Hollanders, that kill'd a white Bear so Gigantic that they found him fourteen foot long.[17]

[BP 39, fols 49–51]

Wood Softened Underground[18]

A curious & intelligent Traveller whose <native> country borders upon that part of the Netherlands <where>[xiv] Massiek is seated told me that near that Town there is a place, that both he and several others went out of curiosity to visit, and that having to satisfy himself, being let down into a great pit that had been made by digging, he found at the depth of about 10 foot some Trees lying in the ground, that there seem'd to be rotten and were said to be very soft. His[xv] diffidence made him remove som of the earth that lay against the side & lower part of one of

---

15   In margin: 'Obs. about the Sp. Ceti Whale'.
16   In margin: 'The Sea Unicorn'.
17   In margin: 'Vast Bears the distance at which they may be met with from the shore'.
18   The identity of Boyle's informant is unclear. By 'Massiek', he probably means Maastricht. The passage appears on its own on a single sheet of paper. It is marked 'Tbd' in the margin by the first line, and the whole has been crossed through in pencil.

those trees, <(but> of what species he answer'd me, He <could> not discover) and having presently[xvi] thrust his Arm as low as he could, he found the wood there so soft that he was able <easily enough> to fill his hand with as much of it as he could grasp, but having kept this obsequious matter a while in the free Air, <it> turn'd to be as hard as ordinary wood.

[BP 37, fol. 119]

Strange Effects of the Tarantula's Bite[19]

A Virtuoso that <was a native of>[xvii] the kingdom of Naples being ask [*sic*] by me wether he had observd any thing about the Terrentula for the persons bitten by it he answerd me that in some part of his or his Fathers estate those insects did abound that he had seen many of them alive some of which were rather of the bigness of scorpions then spiders thô they seem to belong rather to this latter kind[;] as for the poyson they communicate by biting he said he[xviii] long doubted of the truth of the <popular> reports that were believd about it but that he afterwards met with several persons that had been bitten by those venomous insects & that they told him that they were subject to divers odd & sometimes very painful symptoms which when the fits were high would leave them /fol. 76v/ no quiet till[xix] by musick they <were prompted> to dance which they did long & with violence enough till being at length tired they were usualy put into a sweat & fell asleep by which means when they wakend they found themselves very much relievd for that bout. And[xx] my Relator further answerd me that thô they would almost by any brisk musick be manifestly excited to dance & as it were assisted to continue it yet they found themselves much more so by some peculiar tunes which <on>[xxi] that Reason have had a name given them from the Terentula & added that the poysond persons did usualy delight in gay & vivid colours for which Reason they <were> usualy furnisht particularly when they danct with great turfs of Ribbin, blew green red & yellow.[20]

[Bodleian MS Locke c. 37, fols 77v–76v]

---

[19] This anecdote is included with others that were published in Boyle's posthumous *General History of the Air* (1692), the manuscript of the first half of which survives among the manuscripts of its editor, John Locke, in the Bodleian Library, Oxford. For the other anecdotes see *Works*, vol. 12, pp. xx, 43–4 and 128–9. The identity of Boyle's informant is unclear, but, as a resident of the Kingdom of Naples, he must have seemed a good person to ask about the tarantula. A copy of this text in the hand of Locke's amanuensis, Sylvester Brownover, is to be found in Bodleian MS Locke c. 31, fol. 136: this is a clean copy, with differences in spelling and capitalisation that have not here been noted, though a couple of ambiguous words in the original have been elucidated on the basis of Locke's copy.

[20] 'Turf' is an archaic word for a head-dress (*OED*).

Dissolution of Gold

A Traveller very curious & vers'd in Medicinal and Chymical affairs, affirm'd to me having [visited?]ˣˣⁱⁱ within these 2 or 3 dayes that in the East Indys he some while since met with a skilful native of the Country who show'd him the Juice of an Herb or Plant but would not show him the Vegetable itself, Of which Juice he put some <into the hollowd>ˣˣⁱⁱⁱ Palm of his hand & making use of that as a vessel he therein dissolv'd in a short time without the help of the fire a thin piece of Gold & when I askt the Relator whether he had seen this Experiment more than once, he answer'd me that the Brammen [or]ˣˣⁱᵛ Banyan (for I know not which of the two it was) gave some of this Juice to <one> Monsieur a Person of my Relators acquaintance who reiterated the experiment in his presence with the like success, dissolving in the palm of his hand no less a piece of Gold than a Ducket.ˣˣᵛ

²¹Asking within these two dayes an ingenious Acquaintance of the *Liegois*, that has the famous Secret of drawing the *Tinctura auri* without <any> Corrosive Liquor, whether he had seen the Experiment himself he answer'd that his friend had been so kind, as to make it <purposely> in his presence, & whenˣˣᵛⁱ I askt whether he had himself tasted of the menstruum, he told me that he did, and found it, thô not altogether insipid, yet to [have] a very little more taste, which little he judg'd to be acid than common water, and he further answer'd, 1ˢᵗ That the Gold was not prepar'd by Calcination or so much of being brought to fylings, but castˣˣᵛⁱⁱ in an entire piece into the Liquor. 2ˡʸ That the work was done by the help of heat in a few hours as 4 or 6 thô in the Cold, the extraction requires at lest 24 hours. 3ˡʸ that the Artist told him that he had about 4 gr[ains] of tincture out of 1 [dram] of Gold. 4. That the Relator could not perceive the menstruum to work upon the metal like an Aqua Regia upon [Gold] that is with tumult & agitation & store of ascending bubles. 5. That the Tincture extracted was not red, but of a golden color. 6. And that it was not thin but as he exprest it of a spermatic consistence, the impregnated menstruum being ropy almost like Oyl or Hony. 7ˡʸ that the menstruum was drawn out of the air, & that without anyˣˣᵛⁱⁱⁱ Instrument.

[BP 28, p. 277]

---

²¹   A transcribed version of this paragraph appears in BP 25, p. 309. The two versions are identical except for one substantive difference that is noted in n. xxvi; minor differences of orthography have been ignored. In BP 28, p. 277, the text from 'at lest 24 hours' onwards appears in the margin, keyed by a symbol to the end of the text at the bottom of the page. Initially, the continuation was begun at the top of the page, but at 'of Gold' this was discontinued and it was then repeated adjacent to the main entry. Boyle's informant in this paragraph has not been identified, and the same is true of that in the previous one.

'A Note touching the strange, and incredible virtue of some Herbes'

I was told by an Arabian that was here in England, that in a certain Mountain of Barbary called Chis vel Quitz not far from Morocco, grows an Herb of strange vertue, which converts with his juice the imperfect metalls into most pure [gold].[22] It comes about the begining of the Spring, and shines in the night like a Glow-worm.

Sir Thomas Row [*sic*] told me likewise that when he was Embassador in the East Indies, that he became acquainted with one of their Darvis or Holy men, who shewed him the Truth of Alchimy as followeth.[23] First, they prepared [mercury] in this manner. They took a new laid Egge, made a little hole in the top thereof and sucked out the white, then they poured in [1 oz] of [mercury] and clos'd up the hole with clay & salt, this done they laid it to the sun, upon a dry cow-turd, and often turned it all that day, the next morning they poured the [mercury] into a wooden Dish and washed it often with water, so it became very clear & shining bright, then the Darvis cut a dry cow-turd hollow and thereon laid the [1 oz] of prepard [mercury]. then made a little furnace round about it of cow-turds with a few Bricks for them to lean upon, in forme of a Bee-hive, then they gave fire, and cast upon it a little black-powder which raised a red fume, which diminished by degrees until it altogether vanished away, then the Darvis took a Box wherein were dryed Leaves of a certain Herb, and caused the Embassador to crumble with his fingers one of the Leaves into powder, and involving it within the Leave of a Tree that grew hard by them, they cast it upon the [mercury] which instantly boiled, and begun first to congeal at the sides, and so by degrees til all was congealed just after the manner that molten Lead cooles; this done, it was found most excellent silver; when he made Sol, he projected not the black powder upon [mercury] but prepard another powder that was yellow, and then the powder of the Herb. He said moreover that the name of the Herb was Baloune, and that it shined like the moon in the night, and grew in mountains but could only be found but by Holy & good men, and that by the special grace of God; otherwise that thô a man did see it, yet that when he approach'd it, it would vanish out of his sight, He said moreover that it was the great Agent of Nature.

[BP 25, p. 311]

---

[22] The mountain in question was presumably part of the Atlas range, but has not been identified. An accompanying marginal note reads 'Leonardo Fierovanti saith, that Ophioglossone shineth in the night', a reference to one of the books of secrets by Leonardo Fioravanti (1517–1588).

[23] Although the name is clearly written 'Row', this must be a reference to Sir Thomas Rolt (*c.* 1631–1710), who was President of the East India Company's factory at Surat from 1678 to 1682. A further narrative about India supplied by him appears as Workdiary 36–68, where he is described as 'President of all the English in the East Indies'.

Borneo

The President of the English at Surat answer'd me, that he could not by the Enquiry of our Navigators procure any other then a very slender account of the great Island of Borneo; and that the only Trade we drive there, is for Diamonds, of which the Islanders may have great plenty if they would suffer them to be digg'd up. But both the Governor and others of the Court, told me, that those Diamonds were inferior to those of the kingdom of Colchonda, which they take to be the best of India.[24]

[BP 39, fol. 200]

Textual Endnotes

i   Followed by 'I' deleted. In the next sentence, 'where' followed by 'G' deleted; 14 words after that, 'King' followed by 'for' deleted; 19 words after that, 'ounces' preceded by a deleted attempt at the word. At the start of the next paragraph, 'c' is deleted after 'Capital'.

ii   Followed by 'ce' [?] deleted. 11 words later, 'the' followed by a deleted attempt at 'month'; four words after that, '&' is deleted after 'River'; 18 words later, 'Tides' followed by 'tho' deleted, and three words after that, 'Less' replaces 'altogether' deleted.

iii Followed by 'Trade winds' deleted.

iv   Followed by 'pin' deleted; 13 words later, 'wi' deleted after 'place'; 19 words after that, 'qv' deleted after 'puting', and 16 words after that 'paw' deleted before 'patient'. The whole of this paragraph has been lightly crossed through in pencil.

v   Followed by 'of the' deleted; four words later, 'same' deleted after 'the'.

vi   Followed by 'after he had Transiently' deleted; 14 words later, a character is deleted after 'found'.

vii Followed by 'properly so call' deleted; 11 words later, '10' followed by 'Inches' deleted; 12 words after that, 'structure' is altered in composition and followed by a deleted character

viii Replacing 'These ser' deleted. Seven words before this, 'hand' is deleted before 'hand'; 17 words later, a character is deleted after 'by', and the next word, 'the', is followed by 'help of' deleted; 21 words after that, a character is deleted after 'several'. The whole of this paragraph has been lightly crossed through in pencil.

ix   This was subsequently deleted with a view to being replaced by the phrase 'The several times mentiond Priest sent to the King of Siam' (within which 'The' is altered from 'They'), which is inserted in the margin in the hand of Bacon, who has written 'Tb'd' immediately beneath it. At this point, the handwriting of the main text alters from hand C to that of Hugh Greg. The square brackets in the next line are Boyle's.

---

24   For Sir Thomas Rolt, the President, see above, n. 23. The superior diamond mines to which reference is made were those of Golconda. After 'Court' is an insertion mark, but no insertion, as if Boyle intended to identify the Governor and other members of the Court of the East India Company more fully. At the time when Rolt was President of Surat, the Governor was Sir Josiah Child (1630–1699).

x   Followed by 'a couple so' deleted. The whole of this paragraph has been lightly crossed through in pencil.

xi   Followed by 'by' deleted.

xii   Altered in composition. The bottom half of fol. 50 is blank.

xiii   Altered in composition.

xiv   Replacing 'of' deleted; the next word, 'Massiek', is altered from 'Massiete' or 'Massettes'; two words later, 'so that' is deleted; six words after that, 'place' is deleted before 'Town'; 11 words after that, 'who' is deleted after 'others', and six words after that, 'it' is deleted after 'visit'.

xv   Altered from 'He'. 21 words later, 'trees', followed by '(whose kind he ans that' deleted. Eight words after that, 'could' replaces 'was' deleted; the next word, 'not', is followed by 'able to' deleted.

xvi   Followed by 'pres' deleted; 10 words later, 'found' followed by 'that' deleted', and the next word, 'the', followed by 'was ab' deleted.

xvii   Replacing 'had spent much of his life', itself replacing 'had an estate in' deleted; eight words later, 'me' followed by 'about the medicins' deleted; later in the sentence 'estate' followed by 'w' [?] deleted.

xviii   Followed by 'was' deleted. two words later, 'doubted' altered from 'doubtfull'; five words after that, 'the' followed by '<received gener> reports of' deleted; the next word, 'popular', replaces 'general' deleted; nine words after that, 'he' followed by 'found one' deleted, and the next word, 'afterwards', followed by 'w' deleted.

xix   Followed by 'the' deleted; four words later, 'were prompted' replaces 'set a dancing' deleted; six words after that, 'long' followed by 'are' [?] deleted; 10 words after that, 'they' followed by 'usual' [?] deleted, and eight words after that, 'fell' followed by 'as' deleted at end of line.

xx   Followed by 'he fur' deleted; 13 words later, 'brisk' followed by 'le' deleted, and four words after that, 'excited' followed by 'y' deleted.

xxi   Replacing 'for' deleted, which is the reading in MS Locke c. 31, fol. 136; followed by '<those necessary [?] to anothers>' deleted; 10 words later, '&' followed by 'accordingly therefore they would you' deleted; 13 words later, 'colours' altered from 'coloured' [?]; three words after that, 'Reason' followed by 'k' deleted; two words after that 'were' replaces 'had' deleted, and 12 words after that, 'blew' altered from 'blewe'.

xxii   After 'having', 'the' is deleted, but a lacuna is left in the sense which the suggested insertion is intended to fill. In the margin at this point, the word 'Tbd' appears.

xxiii   Replacing 'on the' deleted; the insertion was originally 'the hollowd', but 'the' was deleted and replaced by 'into the'. 19 words later, 'that' is deleted after 'time'; 14 words after that, a character is deleted after 'I'.

xxiv   In the original, 'Brammen' is deleted, but it has been reinstated as the subsequent phrase does not make sense without it; the brackets are accidently closed after 'not', and this has been ignored here. 13 words later, 'one' replaces 'a' deleted, and 18 words after that, 'with' is deleted after 'dissolving'.

xxv   'Asking', the first word of the next paragraph, was originally written here, but was then deleted and a space left before the new paragraph was begun. In the margin adjacent to the start of the new paragraph, the word 'Tbd' appears.

xxvi Followed by 'he as' deleted. In the next line, 'have' has been added for the sense, following its addition in the version in BP 25, p. 309.

xxvii Followed by a deleted character.

xxviii The reading is uncertain as the MS is damaged at this point.

Index